FLORA OF TROPICAL EAST AFRICA

THELYPTERIDACEAE

BERNARD VERDCOURT

Terrestrial ferns. Rhizome creeping or erect with brown non-peltate scales. Stipe not articulated, with 2 vascular strands at base which fuse upwards to form a single U-shaped strand. Lamina 1–2-pinnate or rarely 3-pinnatifid, usually narrowly oblong in outline, glabrous to pubescent or pilose; the indumentum often comprises stiff acicular unicellular hairs, glands of various sorts and fine soft hairs, sometimes hooked at the apex (multicellular hairs present only in one introduced genus, see addendum p. 41); scales are also sometimes present in two African genera along the costae or costules. Veins free or few to many pairs of veins arising from adjoining costules anastomosing into a vein which runs to the sinus between the pinna-lobes. Sori round, with or without a reniform indusium to linear and exindusiate. Spores monolete with perispore.

Holttum divides the family up into 23 genera and stated perhaps 1000 species worldwide. It was partly the large number in Asia (some ten times as many as in Africa) and the cumbrous nature of the Asian species being treated as one genus that forced him to the practical alternative of separate genera; but he admitted that for those concerned only with Africa or part of Africa it was a perfectly reasonable alternative to adopt the one genus *Thelypteris* as was done by Schelpe in F.Z. Pteridophyta. A.R. Smith in Kubitzki, Fam. Gen. Vasc. Pl. 1: 263–272 (1990) and in Fl. Mesoamericana 1, has recognised 5 genera with numerous subgenera. J.P. Roux, Consp. S. Afr. Pterid. 113–122 (2001) follows Holttum's classification. Twelve genera, 26 species and one hybrid are dealt with in the present account. Many of the species have been placed in numerous different genera during the past 250 years and for some species the resulting synonymy runs to several pages. A great deal of this that is not relevant to East Africa has been omitted.

---

* Completed almost entirely from revisions of R.E. Holttum particularly his account of the African species in J. S. Afr. bot. 40: 123–168 (1974), and Malesian species in Fl. Mal., ser. II, 1: 331–599 (1981). His own annotated copy of the 1974 paper has been very useful. I have, however, in a few cases found his species concepts too narrow.

1

5. Veins all free or basal veins sometimes just
     touching below the sinus membrane 6
   At least basal pair of veins anastomosing with
     excurrent vein running to sinus 11
6. Indusium present (but sometimes very small
     and often deciduous – examine many sori) 7
   Indusium absent 10
7. Lower pinnae more or less abruptly reduced
     into a long series of auricles 8
   Lower pinnae gradually reduced or with
     only lowest auriculiform 9
8. Rhizome creeping; indusium hairy ....   4. **Christella** (in part) (p. 10)
   Rhizome erect; indusium glabrous or with
     short glandular hairs ....   3. **Pseudocyclosorus** (p. 8)
9. Indusium well developed ....   4. **Christella** (in part) (p. 10)
   Indusium small and falling early ....   2. **Amauropelta** (in part) (p. 5)
10. Sporangia setose ....   5. **Stegnogramma** (p. 20)
    Sporangia not setose (indusia so small and
      deciduous that may be treated as absent)   2. **Amauropelta** (in part)(p. 5)
11. Rhizome long-creeping; costae usually with
      some scales but can be missing; no reduced
      basal pinnae; usually in swamps and
      marshes ....   7. **Cyclosorus** (in part) (p. 24)
    Characters not as above 12
12. Indusium absent 13
    Indusium present 15
13. Numerous gemmae on rhachis resulting in
      much proliferation; forked hairs present
      (sori forming zig-zag pattern down pinna)   8. **Ampelopteris** (p. 28)
    Only 1 gemma if any on rachis; forked hairs
      absent 14
14. Pinnae narrowly lanceolate, up to 1.5 cm wide,
      strongly narrowed at the base ....   9. **Menisorus** (p. 30)
    Pinnae of mature plants much wider, ±
      truncate at base ....   10. **Pneumatopteris**
                                             (in part) (p. 30)
15. Yellow or whitish capitate hairs abundant on
      distal parts of veins of lower surface; sori
      confined to pinnae-lobes ....   11. **Amphineuron** (p. 37)
    Characters not as above 16
16. Sporangia setose ....   10. **Pneumatopteris**
                                             (in part) (p. 30)
    [some specimens of the hybrid *Chrismatopteris holttumii* with key here – see p. 19]
    Sporangia not setose (but sometimes
      glandular) 17
17. Pinnae deeply divided; either pinnae
      practically devoid of glands but indusia
      densely hairy, or pinnae with round
      yellowish glands or ellipsoid orange or
      reddish glands beneath and indusia with
      small glands ....   4. **Christella** (in part) (p. 10)
    Pinnae subentire to crenate or shortly lobed 18

18. No glands on lamina or indusia, the latter
    glabrous or hairy . . . .   10. **Pneumatopteris**
                                                             (in part) (p. 30)

    Dense yellow or brownish glands on lower
    surface of pinnae and on indusia; pinnae
    distinctly coriaceous in one species . . . . .   12. **Sphaerostephanos** (p. 38)

1. PSEUDOPHEGOPTERIS

Ching in Acta Phytotax. Sinica 8: 313 (1963)

Rhizome erect or suberect (in African species) but elsewhere prostrate or in one species long-creeping; scales thin with short hairs. Stipe and rachis glossy; lamina bipinnate save in two non-African species with pinnules adnate to the pinna-rachis, mostly ± deeply lobed. Veins usually forked, not reaching margin, thickened at the tips. Scales on lower surface of pinna-rachis and rachis few at maturity of frond, those on distal axes reduced to a single row of short cells with reddish cross-walls; hairs unicellular, acicular or capitate or both. Indusium absent; sori round, ± elliptic or rarely elongate. Sporangia sometimes hairy.

Twenty species from St. Helena, São Tomé and Bioko, E to W Tropical Africa, Madagascar, Mascarene Is, mainland Asia, Malesia, Samoa and Hawaii.

Fronds bipinnate-pinnatifid; largest pinnules up to 5.5 cm long;
    pinnule-lobes entire or at the most round toothed; basal pinnules
    mostly not smaller than middle pinnules (widespread) . . . .   1. *P. cruciata*
Fronds bipinnate-bipinnatifid; largest pinnules up to 9.5 cm long;
    pinnule-lobes pinnatifid; basal pinnules mostly reduced (**K** 4) . . .   2. *P. aubertii*

1. **Pseudophegopteris cruciata** (*Willd.*) *Holttum* in Blumea 17: 21 (1969) & in J. S. Afr. Bot. 40: 129 (1974); Schippers in Fern Gaz. 14: 197 (1993); Faden in U.K.W.F. ed. 2: 31 (1994). Type: Mauritius, *Bory* s.n. (B-W 19821, holo., seen by Holttum)

Rhizome erect; fronds tufted, 0.5–1.6 m tall. Stipes reddish or straw-coloured, 30–45 cm long, with numerous spreading hyaline scales at base, turning brown with age. Lamina bipinnate-pinnatifid, 0.3–1 m or more long, 16–58 cm wide; pinnae lanceolate, ± sessile or stalklet to 5 mm, 15–29 cm long, 4.5–9 cm wide; pinnules in ± 25 pairs, 2.5–5.5 cm long, 1.2–1.5 cm wide, distinctly narrowed at the base but always adnate to pinna-rachis, lobed, the lobes oblong, 6–7 mm long, 2.5–3.5 mm wide, entire or almost so or with a few obscure lobes; costa of pinnules pilose; veins simple or forked. Sori on acroscopic vein-branches. Fig. 1, p. 4.

UGANDA. Ankole District: Kalinzu Forest, 4 km NW of saw mill, W of Rubuzigye, 19 Sept. 1969, *Faden et al.* 69/1142!
KENYA. Fort Hall District: road over Kyama R., below Mbugiti School, 13 July 1969, *Faden & Evans* 69/891!; Meru District: NE Mt Kenya, Ithanguni, Kirui volcanic cone, 28 Feb. 1970, *Faden & Evans* 70/86; Kericho District: 4 km SSE of Kericho, Marinyn Tea Estate, along Saosa R., 11 June 1972, *Faden et al.* 72/323!
TANZANIA. Moshi District: Kilimanjaro, Mweka route, 8 Oct. 1973, *Zogg* 115/27!; Morogoro District: NW slope of Bondwa, along road to Morningside, 26 Sept. 1970, *Faden et al.* 70/655!; Iringa District: Udzungwa Mts, W Kilombero Forest Reserve, Nyumbenito Mt, Udekwa Village, Dec. 1981, *Rogers & Hall* 1906!
DISTR. **U** 2; **K** 4, 5; **T** 2, 6, 7; Liberia, Ghana, Cameroon, Congo (Kinshasa); Mascarene Is.
HAB. Swampy stream bank in wet evergreen forest of *Podocarpus, Hagenia* etc.; 1450–2350 m

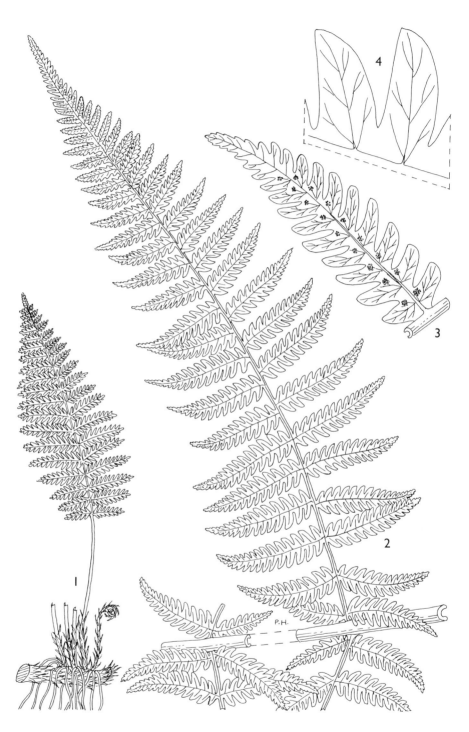

Fig. 1. *PSEUDOPHEGOPTERIS CRUCIATA* — **1**, habit (not drawn to scale); **2**, part of frond ×
²/₃; **3**, fertile pinna × 2; **4**, part of pinna showing venation. All from *Zogg* 115/27. Drawn by
Pat Halliday.

SYN. *Aspidium cruciatum* Willd., Sp. Pl. ed. 4, 5: 278 (1810)
  Thelypteris cruciata (Willd.) Tardieu in Not. Syst. 15: 91 (1954); Alston, Ferns W.T.A.: 61 (1959)
  Macrothelypteris cruciata (Willd.) Pic.Serm. in Webbia 23: 179 (1968)
  M. aubertii sensu Faden in U.K.W.F. ed. 1: 54 (1974), *non* (Desv.) Pic.Serm.

2. **Pseudophegopteris aubertii** (*Desv.*) *Holttum* in Blumea 17: 18 (1969) & in J. S. Afr. Bot. 40: 129 (1974); Faden in U.K.W.F. ed. 2: 31 (1994). Type: Réunion, collector not stated (P) (see Webbia 23, t. 6 (1968))

Very similar indeed to the last species differing essentially only in the characters given in the key. Fronds in material seen up to 1.2 m, lamina bipinnate-bipinnatifid; pinnae up to 40 cm long, 15 cm wide with pinnules up to 9 cm long, 2.5 mm wide, the lobes distinctly pinnatifid; basal pinnules more distinctly reduced.

KENYA. Embu District: Castle Forest Station, 19 Dec. 1972, *Gillett & Holttum* 20097!; District uncertain: S slopes of Mt Kenya, *Pichi Sermolli* 6859 (cited in Webbia 23: 179 (1968))
DISTR. **K** 4; Réunion, Madagascar
HAB. Moist montane valley evergreen forest with *Podocarpus, Dracaena, Cyathea* etc.; ± 2050 m

SYN. *Polypodium aubertii* Desv. in Mem. Soc. Linn. Paris 6: 243 (1827)
  Macrothelypteris aubertii (Desv.) Pic.Serm. in Webbia 23, 177, fig. 2 (1968)

NOTE. Pichi Sermolli discusses the types of the two species in great detail in the reference cited. Holttum overlooked that Pichi Sermolli had collected *P. aubertii* on Mt Kenya.

2. AMAUROPELTA

Kunze, Farnkr. 1: 86, 109, t. 51 (1843)

Rhizome erect (save in one non-African species), the basal part often covered with a mass of tangled roots; scales firm, rather broad with few hairs. Stipe short. Lamina pinnate, gradually narrowed to both base and apex; aerophores* at pinna-bases usually swollen; pinnae deeply lobed. Veins simple except in enlarged basiscopic pinna-lobes; basal veins passing to margin above sinus or connivent just below sinus (in one non-African-mainland species anastomosing with short excurrent vein). Lower surface sometimes with red or orange sessile glands or capitate hairs, or acicular hairs which are sometimes hooked. Sori with very small indusia with glandular or acicular hairs or effectively without indusia.

A genus of about 200 species, six in S & tropical Africa, Madagascar and the Mascarene Is., the rest in America save for three or four in Sri Lanka and Pacific Is. (according to A.R. Smith).

Some hooked hairs always present (in E African specimens)
  on undersurface of pinnae (but careful adjustment of
  microscope focus is needed); no red or orange glands . . . .   1. *A. bergiana*
Hooked hairs absent but reddish sessile or shortly stalked
  glands mostly present (but can be entirely absent) . . . . . .   2. *A. oppositiformis*

1. **Amauropelta bergiana** (*Schltdl.*) *Holttum* in J. S. Afr. Bot. 40: 133 (1974); Schelpe & Diniz, Fl. Moçamb., Pterid.: 206 (1979); W. Jacobsen, Ferns S. Afr.: 382, fig. 286 a & b (1983); Pic. Serm. in B.J.B.B. 53: 275 (1983); Schippers in Fern Gaz. 14: 196 (1993); Faden in U.K.W.F. ed. 2: 31 (1994). Type: South Africa, Kirstenbosch, *Bergius* s.n., *Mundt & Maire* s.n. (Ubi?, syn.)

Rhizome erect with scales up to 8 mm long, sparsely ciliate. Fronds tufted, up to 1–1.2 m tall and 25 cm wide. Stipe grey-brown, 5–17 cm long, minutely pubescent. Lamina narrowly elliptic, 30–95 cm long, up to 25 cm wide, deeply bipinnatifid,

* Aerophore: strictly a pore to facilitate breathing, but here very reduced lower pinnae

acuminate with deeply pinnatifid terminal segment; lower pinnae gradually decrescent; largest pinnae narrowly oblong, 7.5–15 cm long, 1.5–2.5 cm wide, acuminate; lobes oblong, 8–10 mm long, 4 mm wide, obtuse, entire with 8–10 veins, pubescent on costae etc. above, with short hooked hairs beneath (absent in one variety). Indusia very small with few hairs or absent. Fig. 2/1–5, p.7.

var. **bergiana**

Pinnae with hooked hairs present on undersurface of lobes.

KENYA. Meru District: NE Mt Kenya, Ithanguni Forest, road along base of volcanic cone Kirui, 17 km from Nbubu, 22 June 1969, *Faden et al.* 69/768!; Kericho District: 5 km NW of Kericho, Kimugung R., 10 June 1972, *Faden et al.* 72/288!; Teita District: Taita Hills, near Ronge Forest station, *Reichstein et al.* 2917!

TANZANIA. Moshi District: Kilimanjaro, Kibosho to Mt Meru, Dec. 1905, *Daubenberger* 19!; Lushoto District: Soni, 24 Nov. 1973, *Faulkner* 4811!; Ufipa District: Sumbawanga, Chapota, 7 Mar. 1957, *Richards* 8536!

DISTR. **K** 4, 5, 7; **T** 2–4, 6, 7; Cameroon, Bioko, Rwanda, Burundi, Sudan, Ethiopia, Zambia, Malawi, Zimbabwe, South Africa; Mascarene Is.

HAB. Wet evergreen forest, bamboo–*Podocarpus*, usually in boggy areas by streams, rock caves in *Erica* forest; 1200–2500(–?2600) m

SYN. *Polypodium bergianum* Schltdl., Adumbr. Fil. Prom. B. Spei: 20, t. 9 (1825)
  Aspidium maranguense Hieron. in P.O.A. C: 85 (1895). Type: Tanzania, Kilimanjaro, Marangu, *Volkens* 1267 (B, holo.; BM!, iso.)
  Nephrodium bergianum (Schltdl.) Bak., Syn. Fil.: 269 (1867); Hieron., V.E. 2: 11 (1908); F.D.-O.A. 1: 55 (1929)
  Dryopteris bergiana (Schltdl.) Kuntze, Rev. Gen. Pl. 2: 812 (1891); Sims, Ferns S. Afr. ed. 2: 93, t. 10 excl. fig. b & c (1915); Chiov., Racc. Bot. Miss. Consolata Kenya: 139 (1935) pro parte; C. Chr. in Dansk Bot. Ark. 7: 44, t. 9/6–12 (1939)
  Thelypteris bergiana (Schltdl.) Ching in Bull. Fan Mem. Inst. Biol. Bot. 10: 251 (1941); Alston, Ferns W.T.A.: 61 (1959); Tardieu, Fl. Cameroun 3: 240 (1964); Schelpe, F.Z. Pterid.: 193, t. 55/B (1970) & Expl. Hydrobiol. Bassin L. Bangweolo & Luapula 8 (3) Ptérid: 80, fig. 24B (1973) & C.F.A. Pterid.: 150 (1977); Anthony & Schelpe, F.S.A. Pterid.: 219, fig. 71/2 (1986); Burrows, S. Afr. Ferns: 266, t. 44/5, illustr. 62/273, 273a & b (1990)

NOTE. Var. *calva* Holttum without hooked hairs occurs on Mt Cameroon and similar plants in Réunion and the Comoro Is. *Daubenberger* 44 (Kilimanjaro, Kiboscho, 1907) bears the name *Dryopteris bergiana* var. *glanduligera* Rosenstock but I do not know if it was described.

2. **Amauropelta oppositiformis** (*C.Chr.*) *Holttum* in J. S. Afr. Bot. 40: 135 (1974); Pic. Serm. in B.J.B.B. 53: 276 (1983); Jacobsen, Ferns S. Afr.: 384, fig. 287 (1983); Schippers in Fern Gaz. 14: 196 (1993); Faden in U.K.W.F. ed. 2: 31 (1974). Type: Madagascar, Centre, Betafo, *Perrier* 7582 (P, holo., seen by Holttum)

Rhizome erect with narrowly ovate brown scales 4 mm long and tufted fronds 0.3–1.2 m tall. Stipe 2–33 cm long, thinly pubescent. Lamina lanceolate to narrowly oblong-lanceolate, 30–90 cm long, 3.5–21 cm wide, bipinnatifid; lower pinnae descrescent with usually 3–7 lowest pairs much reduced, auriculate; middle pinnae narrowly lanceolate, 2.5–12 cm long, 0.3–1.8 cm wide; lobes oblong to linear-oblong, 8 mm long, 2–3 mm wide with 6–7 pairs of veins, costae hairy on both sides, typically with red or orange glands beneath but these vary greatly in quantity and can be entirely lacking; hooked hairs absent (but see note). Indusium very small and soon deciduous or ± absent. Fig. 2/6–8, p. 7.

UGANDA. Toro District: Ruwenzori, Mobuku Valley, 2 Jan. 1939, *Loveridge* 303!; Kigezi District: Mushongero, Lake Lutamba, 31 Jan. 1939, *Loveridge* 475A!; Bugishu District: Butandiga, 8 Dec. 1938, *A.S. Thomas* 2569!

KENYA. Nakuru District: E Mau Forest Reserve, Camp 10, 1 Sept. 1949, *Maas Geesteranus* 5999!; Kiambu District: near Kinale, Bamboo Forest, 8 Oct. 1959, *Verdcourt & Moggi* 2482!; Meru District: NW Mt Kenya, Marimba Forest, 14 Oct. 1960, *Verdcourt* 2998!

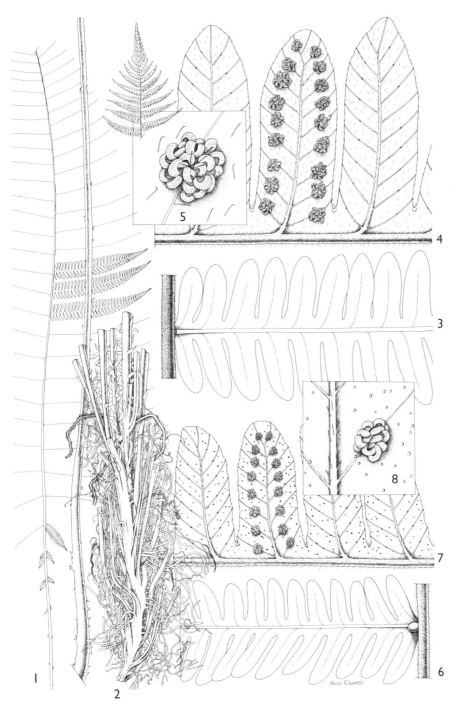

Fig. 2. *AMAUROPELTA BERGIANA* — **1**, habit × ²/₃; **2**, part of rhizome × ²/₃; **3**, part of pinna × 5; **4**, part of pinna showing sori × 5; **5**, undersurface of pinna with sorus detail (enlarged). *AMAUROPELTA OPPOSITIFORMIS* — **6**, part of pinna × 2; **7**, part of pinna showing sori × 4; **8**, undersurface of pinna with sorus detail (enlarged). 1–2 from *Faden & Evans* 69/889; 3–5 from *Faden et al.* 71/900, 72/288; 6–8 from *Zogg* 228/8. Drawn by Ann Farrer.

TANZANIA. Moshi District: Kilimanjaro, Marangu, Himo R., 30 Sept. 1964, *Beesley* 41a; Mpanda
  District: Mahali Mts, Sisaga, 28 Aug. 1958, *Newbould & Jefford* 1836!; Mbeya District: Elton
  Plateau, 30 Nov. 1963, *Richards* 18473!
DISTR. **U** 2, 3; **K** 3, 4; **T** 2, 4, 6, 7; Nigeria, Cameroon, Congo (Kinshasa), Rwanda, Burundi,
  Sudan, Ethiopia, Zimbabwe; Réunion, Madagascar
HAB. Evergreen forest (*Olea – Juniperus* etc.), bamboo forest, *Erica* forest, usually in wet places,
  swamps, near water and open marshy savanna stream banks, also rock ledges on cliff;
  1200–3000 m

SYN. *Dryopteris oppositiformis* C.Chr. in Bonap. Notes Ptérid. 16: 173, t. 2a (1925)
  D. bergiana sensu Chiov. in Racc. Bot. Miss. Consol. Kenya: 139 (1935) quoad *Balbo* 614, *non*
    (Schltdl.) O.Ktze.
  D. squamosa sensu Chiov. in Racc. Bot. Miss. Consol. Kenya: 139 (1935) pro minore parte
    quoad *Balbo* 609, *non* (Schltdl.) Chiov.
  Nephrodium parasiticum sensu Peter, F.D.-O.A. 1: 59 (1929) pro parte, *non* Desv.
  Thelypteris oppositiformis (C.Chr.) Ching in Bull. Fan Mem. Inst. Biol. Bot. 10: 253 (1941);
    Tardieu, Fl. Madag. 5: 272, fig. 37/1–4 (1948); Burrows, S. Afr. Ferns: 270, t. 45/2, illustr.
    62/275, 275a & b (1990)
  T. strigosa sensu Schelpe, F.Z. Pterid.: 193 (1970), *non* (Willd.) Tardieu

NOTE. *Grimshaw* 93/774 (Tanzania, Kilimanjaro, above Lerang'wa (not traced), 2 Oct. 1993)
  has red glands and a few hooked hairs. Whether this is a hybrid needs further
  investigation. Several specimens from Uganda appear to be *A. oppositiformis* or have
  actually been annotated by Holttum as such but lack red glands or have very few. More
  work needs to be done to assess accurately the status of these two taxa. The two occur
  together e.g. *Faden et al.* 762 and 768 both from Kenya, Meru District, Ithanguni Forest. In
  the past most specimens of *A. oppositiformis* were annotated as *A. bergiana*.

3. **PSEUDOCYCLOSORUS**

Ching in Acta Phytotax. Sinica 8: 322 (1963); Holttum & Grimes in K.B. 34:
499–516 (1979)

Rhizome erect or short creeping with thin broad scales with few hairs. Stipe scaly
near base. Fronds simply pinnate, bipinnatifid; basal pinnae ± abruptly reduced and
small or to aerophores with a minute lamina; veins all free with basal aeroscopic vein
passing to base or side of a short sinus-membrane and basal basiscopic vein to other
side of the membrane or to edge above base of sinus; lower surface of pinnae with
some acicular hairs and short or subsessile capitate hairs often present. Sori indusiate.

A genus of 12 species in tropical Africa, Madagascar, Mascarene Is., tropical and subtropical
Asia, Japan, Luzon and Timor.

Pseudocyclosorus pulcher (*Willd.*) *Holttum* in J.S. Afr. Bot. 40: 138 (1974); Holttum
& Grimes in K.B. 34: 513 (1979); Pic. Serm. in B.J.B.B. 53: 277 (1983); Jacobsen,
Ferns S. Afr.: 385, fig. 288 (1983); Schippers in Fern Gaz. 14: 197 (1993); Faden in
U.K.W.F. ed. 2: 31, t. 174 (1994). Type: Réunion, *Bory de St Vincent* 81 (B-W 19787,
holo., seen by Holttum)

Rhizome erect, up to 60 cm tall, often covered with dead frond bases; scales ±
5 mm long. Fronds tufted, 0.8–2 m tall or trailing to 2.5 m. Stipe (measured to first
large pinna) up to 70 cm long, minutely pubescent. Lamina oblong, deeply
bipinnatifid, up to 1.5 m long, 65 cm wide; reduced pinnae up to 10 pairs or more,
lowest 3–5 mm long, uppermost 15 mm, all deeply lobed; largest pinnae 20–33 cm
long, 2–3.5 cm wide, acuminate; lobes narrowly oblong, falcate, up to 1.8 cm long,
4 mm wide, lower surface sparsely hairy on costae and rachis and with short capitate
glandular hairs on costules, veins and lamina; veins up to 18 pairs. Indusium with
short capitate glandular hairs or these ± lacking, sometimes also with acicular hairs.
Fig. 3, p. 9.

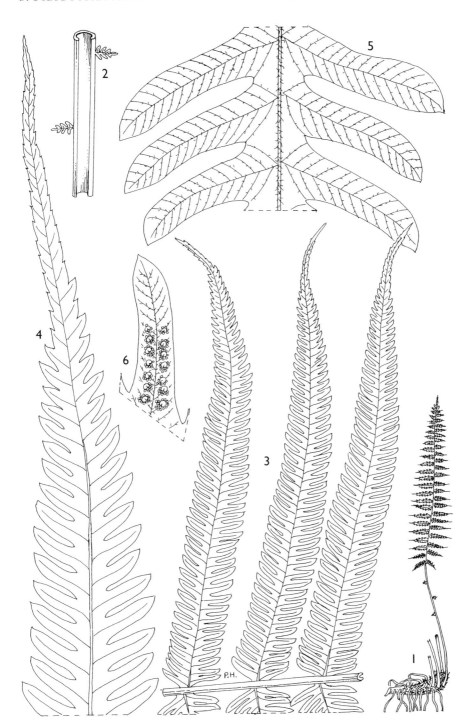

Fig. 3. *PSEUDOCYCLOSORUS PULCHER* — **1**, habit (not drawn to scale); **2**, part of stem showing reduced fronds × ²/₃; part of frond × ²/₃; **4**, single pinna showing venation × 2; **5**, part of pinna showing venation and indumentum (diagrammatic); **6**, fertile pinnule (diagrammatic). 1 from *Nicholson* 46; 2 from *Tweedie* 3226; 3–6 from *Richards* 20849. Drawn by Pat Halliday.

Uganda. Toro District: Bwamba Pass, 16 Nov. 1935, *A.S. Thomas* 1464!; Ankole District: Buhweju, Nyakahanga–Katara, 31 July 1989, *Rwaburindore* 2821!; Mbale District: Mt Elgon at 1800 m, 24 May 1923, *Snowden* 779!

Kenya. Trans-Nzoia District: Mt Elgon at 1800 m, June 1933, *Chater Jack* 19A in CM 5186A!; Fort Hall District: S side of Chania R., between water treatment plant and Chania Falls, 28 Dec. 1968, *Faden* 68/988!; North Kavirondo District: Kakamega Forest, Kibiri Block, S side of Yala R., 21 Jan. 1970, *Faden et al.* 70/25!

Tanzania. Arusha District: Mt Meru, Kykukuma R. gorge, 27 Dec. 1965, *Richards* 20849!; Mpanda District: Mahali Mts, Sabogo, 31 July 1958, *Newbould & Jefford* 1239!; Rungwe District: Tukuyu side of Igali Pass, 27 Aug. 1947, *Brenan & Greenway* 8226!

Distr. **U** 2, 3; **K** 3–5; **T** 2–4, 6–8; Cameroon, Bioko, Congo (Kinshasa), Rwanda, Burundi, Ethiopia, Angola, Malawi, Zimbabwe; Mascarene Is. & Madagascar

Hab. Riverine and swamp forest, seasonally inundated bushland, lake and stream banks and swamp edges; 750–2250 m

Syn. *Aspidium pulchrum* Willd., Sp. Pl. ed. 4, 5: 253 (1810)
> *Nephrodium longicuspe* Bak. in J.L.S. Bot. 16: 202 (1877). Type: Madagascar, *Gilpin* s.n. (K!, holo.)
> *N. zambesiacum* Bak. in Ann. Bot. 5: 318 (1891). Type: Malawi, Shire Highlands, *Buchanan* s.n. (K!, holo.)
> *Aspidium zambesiacum* (Bak.) Hieron. in P.O.A. C: 85 (1895)
> *Dryopteris zambesiaca* (Bak.) C.Chr., Ind. Fil.: 318 (1905); Sim, Fern S. Afr. ed. 2: 103, t. 26 (1915); C. Chr. in Dansk. Bot. Ark. 7: 46, t. 9/13–18 (1932)
> *Nephrodium prolixum* sensu Peter, F.D.-O.A. 1: 56 (1929) quoad *Peter* 15864, *non* Bak
> *Thelypteris zambesiaca* (Bak.) Tardieu in Not. Syst. 14: 345 (1952) & Fl. Madag. 5: 278, fig. 39/4–7 (1958); Alston, Ferns W.T.A. 61 (1959); Tardieu, Fl. Cameroun 3: 243 (1964)
> *T. longicuspis* (Bak.) Schelpe in J.S. Afr. Bot. 31: 262 (1965) & F.Z. Pterid.: 192, t. 55/A (1970)
> *T. pulchra* (Willd.) Schelpe, C.F.A., Pterid.: 151 (1977); Schelpe & Anthony, F.S.A., Pterid.: 215 (1986); Burrows, S. Afr. Ferns: 261, t. 43.3, illustr. 62/264, 264a & b (1990)

Note. Holttum (in J.S. Afr. Bot. 40: 137 (1974)) described *Pseudocyclosorus camerounensis* from Cameroon (Type: Cameroon, Djuttitsa, *Meurillon* CNAC 726 (K!, holo.; P, iso.) characterised by the reduced pinnae, each consisting of a prominent aerophore with no distinct lamina and allying it with a Chinese and Indian species. Later Holttum and Grimes (K.B. 34: 503 (1979)) referred a Uganda specimen to this species (*Loveridge* 306, K!), Ruwenzori, Mobuku Valley at 1850 m, 2 Jan. 1939, beside river) saying it differed in having pinnae 3 cm wide and hairs 1 mm long on lower surface of costae and might be a "distinct local variety". I have been unable to distinguish this from many other Ugandan, Kenyan and Tanzanian sheets of *P. pulcher* many identified as such by Holttum (or as one of its synonyms). There seems continuous variation between small reduced pinnae, aerophores with a small lamina and aerophores. Only study of populations in the field will show if there are constant differences. I certainly do not believe *Loveridge* 306 is closer to a Chinese species *P. tylodes* (Kunze) Ching than to the bulk of the material treated as *P. pulcher*. Many herbarium specimens are devoid of stipes or show only a portion, so these would be impossible to distinguish anyway. The Cameroon material of *P. camerounensis* needs further comparison with West African *P. pulcher*.

4. CHRISTELLA

Léveillé, Fl. de Kouy-tchéou: 472 (1915); Holttum in Taxon 20: 533 (1971) & in Blumea 19: 43 (1971) & in K.B. 31: 293 (1976)

Rhizome erect, suberect or creeping, in some species widely so; scales narrow, ± densely hairy. Lamina simply pinnate, pinnatifid in almost all species with 1–5(–10) pairs of lower pinnae gradually decrescent; lower surface with erect acicular hairs and small capitate hairs and thick orange-red glandular hairs sometimes present. In most species the basal veins or sometimes several pairs anastomose but in a few species the veins are free. Sori indusiate (save in two non-African species); elongate glands present on stalks of sporangia in all but a few species.

About 70 species in tropics and subtropics but the position of some African and Neotropical species needs further work, A.R. Smith treats *Christella* as a synonym of one of his subgenera of *Cyclosorus*. The species are extremely difficult and in need of much further field work.

1. Lowest pair (or two pairs) of veins mostly anastomosing with an excurrent vein to the sinus clearly distinct (Fig. 4) (Sect. *Christella*) 2

   Veins totally free or basal veins connivent or touching at sinus or base of sinus membrane or only a very few anastomoses 9

2. Two pairs of veins anastomosing ...... see hybrids on p. 19 and note on p. 20

   Only one pair of veins anastomosing 3

3. Indusium with dense glands and often hairs as well but glands usually preponderating (veins typically free but anastomosing in some East African specimens) .... 4. *C. microbasis*

   Indusium hairy and sometimes with few glands to entirely glabrous 4

4. Indusium glabrous .... 3. forms of *C. dentata*

   Indusium hairy and sometimes with few glands 5

5. Pinna-lobes with distinct glands on costules 6

   Pinna-lobes without distinct glands but small capitate hairs often present 7

6. Large elongate orange glands on veins of lower surface of pinna-lobes (**U** 2) .... 1. *C. parasitica*

   Dense very pale yellow small glands on lower surface of pinna-lobes .... see 8. *C.* sp. A

7. Indusium with very dense white hairs ± 1 mm long exceeding width of indusium (**T** 3, Usambaras) .... 9. *C.* sp. B

   Indusium densely hairy but hairs relatively shorter compared with width of indusium 8

8. Rhizome erect with fronds tufted; basal veins anastomosing to form a broad low triangle with costa but variable and often some connivent only at base of sinus (widespread) ... 2. *C. hispidula*

   Rhizome short-creeping with fronds shortly spaced; basal veins anastomosing to form a usually more acute triangle (but much variation) .... 3. *C. dentata*

9. Pinnae-lobes with dense small very pale yellowish glands beneath 10

   Pinnae-lobes without such dense small glands 11

10. Indusia densely hairy (**K** 1, ?4, ?6) .... see 9. *C.* sp. A

    Indusia with few to many small glands .... 6. *C. guineensis*

11. Indusium with numerous glandular hairs but usually some acicular hairs as well (lamina often with many lower pairs of pinnae much reduced) 12

    Indusium usually densely hairy with few glands or less often almost glabrous (lamina with no or few reduced pinnae) 13

12. Indusium regularly round and distinct (veins free or anastomosing) .... 4. *C. microbasis*

    Indusium mostly rather small and ± irregular (veins always entirely free) .... 5. *C. friesii*

13. Rhizome erect; fronds tufted .... 6. *C. gueinziana*

    Rhizome creeping; fronds ± spaced .... 7. *C. chaseana*

1. **Christella parasitica** (*L.*) *Léveillé*, Fl. Kouy-tchéou: 475 (1915); Holttum in J.S. Afr. Bot. 40: 140 (1974) & in K.B. 31: 309 (1976) & in Fl. Males. ser. II, 1: 559 (1981). Type: China, Canton, *Osbeck* s.n. (S-PA, Herb. Swartz!, holo., seen by Holttum)

Rhizome short- to long-creeping; scales lanceolate, 7 mm long. Fronds 0.5–1 m tall. Stipe up to 50 cm long, softly hairy. Lamina simply pinnate-pinnatifid, 20–50 cm long with about 20 pairs of pinnae, the basal ones deflexed, not or only slightly reduced; largest pinnae 5–16 cm long, 1–2 cm wide, acuminate, deeply lobed; lobes oblong, 6 mm long, 2.5–4 mm wide, rounded truncate, ± entire; veins in 8–10 pairs, the lowest pair joined forming an excurrent vein to the sinus-membrane, lower surfaces of lobes typically covered with soft spreading hairs and thick orange or very pale ellipsoid glands usually present on the costules and veins. Indusia hairy.

UGANDA. Mengo District: Kyagwe, 10 km W of Lugazi, Ssezzibwa Falls, 8 Sept. 1969, *Faden* 69/985! & Kyadondo, Bukoto, 29 June 1988, *Rwaburindore* 2647! & Kyadondo, near Kisaasi, 13 Apr. 1990, *Rwaburindore* 2976!
DISTR. **U** 4; Congo (Kinshasa); widespread in tropical Asia and Pacific, also St. Helena (introduced ?)
HAB. Forest, seasonally inundated bushland, thicket; ± 1200 m

SYN. *Polypodium parasiticum* L., Sp. Pl.: 1090 (1753); C. Chr. in Ark. for Bot. 9 (11): 26, fig. 4 (1910)
    Cyclosorus parasiticus (L.) Farwell in Am. Midl. Nat. 12: 259 (1931)
    Thelypteris parasitica (L.) Fosberg in Occ. Pap. Bishop Mus. 23: 30 (1962)
    T. fadenii Fosberg & Sachet in Smiths. Contr. Bot. 8: 9 (1972). Type: Uganda, Mengo District: Kampala, Makerere Hill, *Lye* 5186 (MHU, holo.; B, BM!, EA!, G, GH, K!, MO, NY, P, UC, US, iso.)

NOTE. African material is less hairy than typical in Asia. Holttum (1972, 1981) mentions Kenya in his distribution but I have not seen the material. Schippers in Fern Gaz. 14: 196 (1993) mentions that *URFRP* 87430 from Usambaras, Lutindi, and *De Boer & Bosch* 315 from "Bugoba" may belong to this species. In 1971 Fosberg (in litt.) maintained that he still thought *T. fadenii* was a good species but a label on *Faden et al.* 69/998(A) of Fosberg's says possibly too close to *T. parasitica*.

2. **Christella hispidula** (*Decne.*) *Holttum* in K.B. 31: 312 (1976); Schelpe & Diniz, Fl. Moçamb., Pterid.: 208 (1979); Schippers in Fern Gaz. 14: 196 (1993); Faden in U.K.W.F. ed. 2: 31 (1994). Type: Timor, *Guichenot* s.n. (P, holo.)

Rhizome erect with brown to dark brown lanceolate entire or ciliate often thinly pilose scales up to 6 mm long. Fronds tufted, 0.3–1 m tall. Stipe 12–40(–60) cm long, thinly pubescent with minute whitish hairs and scaly at base. Lamina lanceolate, deeply 2-pinnatifid, 25–60 cm long, ± 30 cm wide, apex acuminate with deeply pinnatifid terminal segment; lower 2–4 pair of pinnae deflexed and somewhat reduced; middle pinnae 8–16 cm long, 1.2–2.2 cm wide, narrowly oblong, acuminate, deeply pinnatifid; lobes narrowly oblong, 4–10 mm long, 2–5 mm wide, somewhat falcate, entire, obtuse, hairy on both surfaces and sometimes with small capitate hairs; veins 7–8 pairs, the basal pair anastomosing at a very obtuse angle below the sinus with next pair to edge of sinus. Indusium hairy, the hairs sometimes very long.

UGANDA. Bunyoro District: Budongo, Siba Forest, Jan. 1936, *Sangster* 23!; Bugosa District: close to White Nile, 16 km NW of Jinja, Kibibi, 6 Feb. 1953, *Wood* 624!; Mengo District: Buganda, 14 Feb. 1934, *Longfield* 12! & 15!
KENYA. Kavirondo District: Kakamega Forest, along Yala R., about 5 km SE of Forest Station, 25 Nov. 1969, *Faden et al.* 69/2012!; Kwale District: Shimba Hills, Kurumuji and Mwacho Mwana R. valleys, 2 Aug. 1970, *Faden & Evans* 70/417! & Shimba Hills, near Sheldrick's Falls, 2 Apr. 1968, *Magogo & Glover* 621!
TANZANIA. Moshi District: by R. Njoro, 8 km W of Moshi, 3 Nov. 1955, *Milne-Redhead & Taylor* 7030!; Lushoto District: Amani, 9 Oct. 1929, *Greenway* 1767!; Mpanda District: Kungwe-Mahali Peninsula, Kabwe R. just S of Pasagulu, 7 Aug. 1959, *Harley* 9193!
DISTR. **U** 2–4; **K** 4, 5, 7; **T** 2–4, 6, 7, ?8; West Africa, Bioko, Cameroon, Sudan, Malawi, Mozambique, Zimbabwe; Seychelles, Mascarene Is., Madagascar; widespread in tropical Asia and America

HAB. Evergreen rain forest, often riverine, and on rocks near falls; 150–2400 m

SYN. *Aspidium hispidulum* Decne. in Nouv. Ann. Mus. Hist. Nat. Paris 3: 346 (1834)
   Nephrodium hilsenbergii Presl, Epim. Bot.: 47 (1851). Type: Mauritius, *Sieber* 49 (PRC, holo.;
      P, K!, iso.)
   N. quadrangulare Fée, Gen. Fil.: 308 (1852). Type: Guyana, *Leprieur* 182 (?P, holo.; NY, iso.)
   N. molle sensu Bak. in Trans. Linn. Soc. Lond. Ser. II, 2: 354 (1887) pro parte, *non* (Sw.)
      R.Br.
   Cyclosorus quadrangularis (Fée) Tardieu in Notul. Syst. 14: 345 (1952) & in Mém. I.F.A.N.
      28: 122, t. 21/10–12 (1953); Alston, Ferns W.T.A.: 62 (1959)
   Thelypteris quadrangularis (Fée) Schelpe in J. S. Afr. Bot. 30: 196, t. 1, fig. b (1964) & in F.Z.
      Pterid.: 195 (1970)
   T. hispidula (Decne.) Reed in Phytologia 17: 283 (1968); Burrows, S. Afr. Ferns: 264, t.
      44/2, illustr. 64/268, 268a & b (1990)
   Christella hilsenbergii (Presl) Holttum in J. S. Afr. Bot. 40: 142 (1974); Jacobsen, Ferns S. Afr.:
      387. fig. 289 (1983)

NOTE. *R.B. & A.J. Faden* 74/369 (Lushoto District: E Usambaras, Derema, Hunga R., near small
   falls, 30 Mar. 1974) and other specimens show considerable variation in the anastomosing of
   the basal veins, some joining and having an excurrent vein to the sinus others quite free but
   close at base of sinus (all on one pinna). This is, I feel, just normal variation rather than
   hybridisation which is very often a facile explanation based on no evidence. Some specimens
   I have merely annotated as *C. hispidula/gueinziana* intermediates.

   3. **Christella dentata** (*Forssk.*) *Brownsey & Jermy* in Brit. Fern Gaz. 10: 338 (1973);
Holttum in J.S. Afr. Bot. 40: 143 (1974) & in K.B. 31: 314 (1976); Schelpe & Diniz,
Fl. Moçamb. Pterid. 209 (1979); Holttum, Fl. Males. Ser. II, 1: 557 (1981); Pic. Serm.
in B.J.B.B. 53: 280 (1983); Jacobsen, Ferns S. Afr.: 388, fig. 290, 291 (1983); Schippers
in Fern Gaz. 14: 196 (1993); Faden in U.K.W.F. ed. 2: 32, t. 174 (1994); Hepper &
Friis, Pl. Forssk. Fl. Aegypt.-Arab.: 290 (1994). Type: Yemen, Bolghose, *Forsskål* 809
(C, holo.)

   Rhizome shortly to distinctly creeping, ± 7 mm in diameter, with dark brown ovate
to lanceolate entire thinly pilose scales up to 6–8 mm long. Fronds closely spaced,
0.4–1.5(–2) m tall. Stipe 8–50 cm long, glabrous or slightly pubescent, with scales at
base. Lamina pinnate, elliptic to narrowly elliptic in outline, 0.3–1.3 m long, up to
40 cm wide, acuminate with a deeply pinnatifid terminal segment; lowest 2–4 pairs
of pinnae usually gradually decreasing; middle pinnae narrowly oblong or
lanceolate, (4.5–)8–21 cm long, 1.5–2.5 cm wide with long narrowly acuminate
crenate apex, deeply pinnatifid into oblong lobes 4–10 mm long, 2.5–4.5 mm wide,
entire, obtuse to acute, pilose along the costa above and shortly pubescent beneath;
veins 8–9 pairs with one pair anastomosing with excurrent vein to sinus, the triangle
formed usually acute or in some specimens (including from type area) with 2 pairs
of veins anastomosing (see note after hybrids, p. 20). Indusia with short white hairs
or almost or quite glabrous (see note).

UGANDA. Toro District: Kibale Forest, 16 July 1938, *A.S. Thomas* 2301!; Busoga District: 19 km
   N of Jinja on Kamuli road, Kagoma Local Forest Reserve, 23 July 1953, *G.H.S. Wood* 987!;
   Mengo District: Kampala, Feb. 1939, *Chandler* 2736!
KENYA. West Suk District: Sigor–Kapenguria road at Kipros R. crossing, 15 Mar. 1977, *Faden &
   Faden* 77/800!; North Kavirondo District: Kakamega Forest, NE of Forest Station, 24 Nov.
   1969, *Faden et al.* 69/1979!; Mombasa, Mar. 1876, *Hildebrandt* 1959!
TANZANIA. Mbulu District: Lake Manyara National Park, near the main gate, 15 Nov. 1963,
   Greenway & Kirrika 11021!; Lushoto District: Amani, 14 Sept. 1929, *Glynne* 228!; Mpanda
   District: Mahali Mts, Utahya, 21 Aug. 1958, *Newbould & Jefford* 1699!; Pemba, Mkoani, 8 Sept.
   1946, *R.O. Williams* s.n.!
DISTR. U 2–4; K 2–5, 7; T 1–4, 6, 8; Z; P; throughout the Old World tropics and subtropics, just
   reaches into SW Spain; introduced in some parts of tropical America
HAB. Evergreen forest, often riverine, ditch-sides in swampy areas, valley bushland and thicket;
   (?0–)45–2100 m

SYN. *Polypodium dentatum* Forssk., Fl. Aegypt.-Arab.: 18 (1775)
    P. molle Jacq., Collect. Bot. 3: 188 (1789), *non* Schreb. (1771) *nec* All. (1785), *nom illegit.*
        Type: cult. in Vienna, *Jacquin* s.n. (W, holo.)
    Aspidium molle Sw. in J. Bot. (Schrad.) 1800, 2: 34 (1801), *nom. nov.* pro *P. molle* Jacq.
    Nephrodium molle (Sw.) R.Br., Prodr. Fl. New Holl.: 149 (1810); Bak. in Oliv. in Trans. Linn.
        Soc., Ser. II, Bot. 2: 354 (1887) pro parte
    N. parasiticum sensu Peter, F.D.-O.A. 1: 58 (1929) pro parte, *non* (L.) Desv.
    Dryopteris gongylodes sensu Chiov., Racc. Bot. Miss. Consolata Kenia: 140 (1935) quoad *Balbo*
        621 pro parte, 618 & 623, *non* (Schkuhr) Link
    Thelypteris dentata (Forssk.) E.St.John in Amer. Fern Journ. 26: 44 (1936); Schelpe, F.Z.
        Pterid.: 197 (1970) & in Expl. Hydrobiol. Bassin L. Bangweulu & Luapula 8 (3) Ptérid.:
        81 (1973) & C.F.A. Pterid. 152 (1977); Schelpe & Anthony, F.S.A., Pterid.: 215, fig. 72
        (1986); Burrows, S. Afr. Ferns 263, t. 44/3, illustr. 267, 267a & b (1990)
    Cyclosorus dentatus (Forssk.) Ching in Bull. Fan. Mem. Inst. Biol. Bot. 8: 206 (1938); Tardieu
        in Mém I.F.A.N. 28: 121, t. 21/7–9 (1953); Alston, Ferns W.T.A.: 62 (1959); Tardieu, Fl.
        Cameroun: 248, t. 37/4–5 (1964) & Fl. Gabon 8: 150, t. 24/4–5 (1964)
    C. elatus vel sp. aff. sensu Haerdi in Acta Tropica Suppl. 8: 33 (1964)

NOTE. Holttum states after describing *C. modesta* (in J. S. Afr. Bot. 40: 144 (1974)) "No other
    species of sect. *Christella* has glabrous indusia", but they can be in *C. dentata* and are in fact in
    a number of specimens determined by Holttum. I have not seen the single specimen *C.
    modesta* is based on.
        Plaizier & Breteler 1124 (Kenya, Kisii District, 4 km SE of Marani, 9 Apr. 1978) has the venation
    and rhizome of *C. dentata* but there are glands on the indusia as well as dense hairs and some
    glands on the pinnae. *Allan* 3698 (Uganda, Toro District, Mwamba forest, 5 Dec. 1957) has
    fertile fronds 30 cm long or less and largest pinnae only 3.5 cm long, quite shallowly lobed; the
    creeping rhizome, anastomosing veins and indusium all suggest it is a small form of *C. dentata*.
    Holttum and others have thought that material resembling *C. dentata* but with two pairs of veins
    anastomosing indicated it was of hybrid origin so I was surprised to find that Forsskål's type and
    other material from the Yemen has some lobes with two pairs of veins anastomosing (see p. 19
    under hybrids). Some specimens, e.g. *Sangster* 109 (Uganda, Budongo Forest, Mar. 1931) have
    quite glabrous indusia depite the comment by Holttum (J. S. Afr. Bot. 40: 144 (1974))

4. **Christella microbasis** (*Bak.*) *Holttum* in J. S. Afr. Bot. 40: 146 (1974); Jacobsen,
Ferns S. Afr.: 390, fig. 293 (1983); Schippers in Fern Gaz. 14: 196 (1993). Type:
Nigeria, Onitsha, *Barter* 571 (K!, holo.)

Rhizome erect or short-creeping with brown lanceolate scales 2–6 mm long.
Fronds tufted, 0.3–1.5 m tall. Stipe straw-coloured, 15–60 cm long. Lamina elliptic-
lanceolate, 23–70 cm long, 10–25 cm wide, with 2–3(–11) pairs of lower pinnae
reduced with lowest 2.5 cm long, but sometimes little reduced; largest pinnae
narrowly oblong, 5–16 cm long, 1–1.8 cm wide, acuminate, pinnatifid into oblong
lobes, 5–10 mm long, 2–4 mm wide, falcate, with hairs and short capitate hairs on
both surfaces; veins 8–10 pairs, typically free but in E Africa often with many
anastomosing basal veins. Indusia with numerous glandular hairs and sometimes a
few acicular hairs but glands usually outnumbering the hairs, or if not, then
indusium much rounder and larger than in next species.

KENYA. Meru District: Meru–Reja, 23 July 1912, *Balbo* 811!
TANZANIA. Arusha District: Arusha National Park, Maji ya Chai R., 1 Apr. 1971, *Richards* 26895!
    & same park, Mt Meru Crater slope, Murekamba stream, 27 Dec. 1966, *Vesey Fitzgerald* 5039!;
    Songea District: about 32 km E of Songea, by R. Mkurira, 25 Dec. 1955, *Milne-Redhead &
    Taylor* 7743C!
DISTR. **K** 4; **T** 2, 8; Sierra Leone, Mali, Guinea, Liberia, Ivory Coast, Ghana, Benin, Nigeria,
    Cameroon, Sudan, Angola, South Africa (Transvaal)
HAB. By streams in evergreen forest and rocky ground in gorge near waterfall,
    Brachstegia–Uapaca woodland; 1000–2100 m

SYN. *Nephrodium microbasis* Bak., Syn. Fil. ed. 2: 496 (1874)
    Dryopteris dentata sensu Chiov., Racc. Bot. Miss. Consolata Kenia: 139 (1935) quoad *Balbo*
        811, *non* (Forssk.) C.Chr.

Thelypteris microbasis (Bak.) Tardieu in Mém I.F.A.N. 28: 117, t. 20/1–4 (1953); Alston, Ferns W.T.A.: 61 (1959); Tardieu, Fl. Cameroun 3: 242 (1964); Burrows, S. Afr. Ferns: 265, illustr. 62/269, 269a & b (1990)

NOTE. The *Balbo* specimen was annotated by Faden as *Cyclosorus dentatus* in 1970 but Holttum in 1973 determined it as probably *Christella microbasis*. Faden (U.K.W.F. ed. 2: 31 (1994)) said the record of *C. microbasis* in Kenya by Holttum 1974 is apparently erroneous but the record is clearly based on the *Balbo* specimen. *Bigger* 2138 (Kilimanjaro, Lemosho R., 18 Aug. 1968) & 2327 (Kilimanjaro, Himo R., 17 Nov. 1968) have equal numbers of hairs and glands on the indusia which are larger and more regularly round than in *C. friesii* and the rhizome appears erect but there are up to 11 pairs of reduced pinnae. The relationships of these two taxa need investigation.

5. **Christella friesii** (*Brause*) *Holttum* in J. S. Afr. Bot. 40: 145 (1974); Jacobsen, Ferns S. Afr.: 389, fig. 292 (1983); Schippers in Fern Gaz. 14: 196 (1993); Faden in U.K.W.F. ed. 2: 32 (1994). Type: Zambia, Luvingo, *R.E. Fries* 1104 (UPS, holo.; BM!, photo.)

Rhizome creeping with dark brown lanceolate acute entire scales up to 4.5 mm long. Fronds spaced 1–4 cm apart, 1–1.8 m tall. Stipe 24–38 cm long, greyish brown, thinly pubescent with short white hairs. Lamina deeply bipinnatifid, narrowly oblong-lanceolate in outline, 0.5–1.2 m long, 12–26 cm wide, acute with up to 6 pairs of basal pinnae suddenly decrescent, widely spaced and reduced to less than 2 cm long; pinnae often folded longitudinally when dry, middle narrowly oblong, 6.5–13 cm long, 1.2–3 cm wide, deeply pinnatifid; lobes narrowly oblong, 0.6–1.5 cm long, 2.5–3 mm wide, falcate, acute, costa with dense white hairs above, lower surface pilose with soft white hairs; veins up to 13 pairs, not anastomosing. Indusium often small and ± irregular, glandular and with few to many long white hairs or entirely lacking. Fig. 4, p.16.

KENYA. Trans-Nzoia District: 7.5 km NE of Kitale, Koitobos R., Sandum's Bridge, July 1964, *Tweedie* 2863! & same locality, Nov. 1965, *Tweedie* 3220! & same locality, 12 June 1971, *Faden et al.* 71/458!
TANZANIA. Ufipa District: Chapota, 4 Dec. 1949, *Bullock* 2006!; Mbeya District: Mbosi Circle, Judyland Farm, 10 Jan. 1961, *Richards* 13801!
DISTR. **K** 3; **T** 4, 7; Cameroon, Congo (Kinshasa), Zambia, Malawi, Zimbabwe
HAB. Swamps with sedges, *Eulophia*, *Lobelia* etc., often along forest streams; 1400–1800 m

SYN. *Dryopteris friesii* Brause in Wiss. Ergebn. Schwed. Rhod.-Congo Exp., Bot. 1: 1 (1914)
    Thelypteris friesii (Brause) Schelpe in Bol. Soc. Brot. Sér. 2, 41: 216 (1967) & F.Z. Pterid.: 192 (1970) & Expl. Hydrobiol. Bassin L. Bangweolo & Luapula 8 (3) Ptérid.: 79 (1973); Burrows, S. Afr. Ferns: 260, t. 43/5, illustr. 63/263, 263a & b (1990)

NOTE. It is possible to confuse this with *Amauropelta oppositiformis* which sometimes lacks the characteristic glands. Schelpe gives the lamina length up to 1.7–0.48 m (error for 1.7 × 0.48 m) but although the species appears to be larger in the F.Z. area, I have seen nothing of this size.

6. **Christella guineensis** (*Christ*) *Holttum* in J. S. Afr. Bot. 40: 145 (1974). Type: Guinea, Labè, *Chevalier* 12385 (P, holo.; BM!, photo)

Caudex short-creeping or ± erect with fronds densely tufted. Stipe 25–56 cm long, glabrous or with minute capitate hairs. Lamina narrowly oblong, 30–45 cm long, 15–20 cm wide, with 1–4 pairs of very reduced pinnae, 3–10 mm long (these have been ignored in calculating stipe and lamina lengths); pinnae narrowly oblong, up to 12 cm long, 1.5 cm wide, acuminate, very deeply pinnatifid almost to costa; lobes narrowly oblong to lanceolate, 6–10 mm long, 1.5–2.2 mm wide, straight or somewhat falcate, rounded to acute at apex, with dense small yellowish glands beneath; costules ± 2.5 mm apart; veins not anastomosing; sinus membranes thickened. Indusia small with very few to many small yellowish glands.

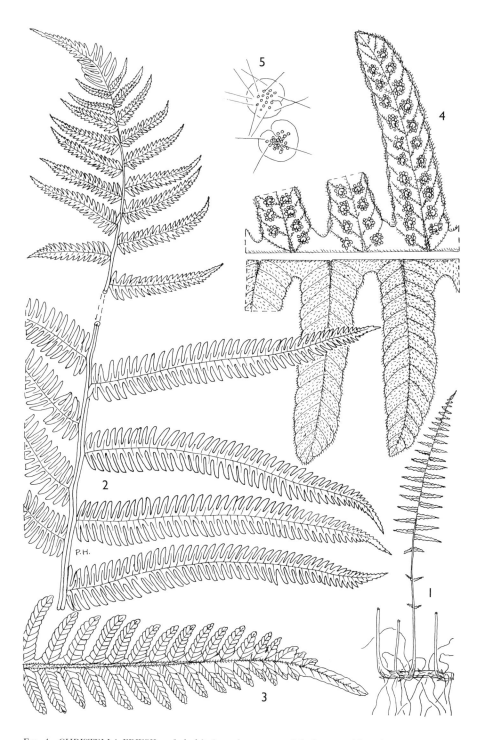

FIG. 4. *CHRISTELLA FRIESII* — **1**, habit (not drawn to scale); **2**, part of fronds × ²/₃; **3**, single pinna × 2; **4**, part of fertile pinna showing indumentum on both surfaces (diagrammatic); **5**, indusia (diagrammatic). All from *Faden et al.* 71/458. Drawn by Pat Halliday.

UGANDA. Toro District: Fort Portal, Nyakasura School, July 1938, *Thompson* 71!; frontier between Toro and Congo (Kinshasa), 24 June 1938, *Thompson* 65!
DISTR. **U** 2; Sierra Leone, Guinea, Nigeria, Cameroon, Angola
HAB. River banks; ?900–1650 m

SYN. *Dryopteris guineensis* Christ in J. de Bot. 22: 22 (1909)
    Thelypteris guineensis (Christ) Alston in Bull. Brit. Mus. Bot. 1: 48 (1952) & in Ferns W.T.A. 61 (1959); Schelpe, C.F.A., Pterid.: 153 (1977)

NOTE. In his key, Schelpe states 'indusios pilosos', so some doubt must rest on his records from Angola. The specimens cited above had 'guineensis' written on in pencil in B. Parris' handwriting and I am sure this is correct.

7. **Christella gueinziana** (*Mett.*) *Holttum* in J. S. Afr. Bot. 40: 147 (1974); Schelpe & Diniz, Fl. Moçamb., Pterid.: 210 (1979); Pic. Serm. in B.J.B.B. 53: 280 (1983) (as *gueintziana*); Jacobsen, Ferns S. Afr.: 391, fig. 294, 295 (1983); Schippers in Fern Gaz. 14: 196 (1993); Faden in U.K.W.F. ed. 2: 32 (1994). Type: South Africa, Natal, *Gueinzius* s.n. (B, holo.)

Rhizome erect with brown, ovate, acute entire scales up to 4 mm long. Fronds tufted, 0.5–1.5 m tall. Stipe straw-coloured, 30–50 cm long, thinly pubescent with minute white hairs. Lamina elliptic in outline, deeply bipinnatifid, 34–90 cm long 18 cm wide with acute deeply pinnatifid apex, basal pair of pinnae somewhat reduced and deflexed; middle pinnae narrowly oblong, 12–20 cm long, 1.4–1.7 cm wide, long attenuate with long narrow crenate segment at the apex, deeply pinnatifid into narrowly oblong falcate acute lobes 4–12 mm long, 2–3.5 mm wide, pilose on both surfaces and often with small glandular capitate hairs beneath; veins 8–11(–15) pairs, the basal pair free or meeting at the sinus membrane. Indusium 0.7 mm in diameter, glandular and pilose with white hairs or sometimes almost glabrous.

KENYA. Trans-Nzoia District: near Kitale, May 1962, *Tweedie* 2353!; North Kavirondo District: Kakamega Forest, Kibiri Block S side of Yala R., 21 Jan. 1970, *Faden et al.* 70/21!; Masai District: Kilgoris, R. Romosha, 8 Sept. 1061, *Glover et al.* 2608!
TANZANIA. Arusha District: Meru Mt, 13 Mar. 1971, *Richards* 26759!; Lushoto District: W Usambaras, 6.4 km NE of Lushoto, Mkuzi, 21 Apr. 1953, *Drummond & Hemsley* 2168!; Iringa District: Mufindi, 4 Apr. 1970, *Paget-Wilkes* 823!
DISTR. **K** 1?, 3–7; **T** 1?, 2–4, 6, 7; Sudan, Ethiopia, Angola, Malawi, Mozambique, Zimbabwe, South Africa; Madagascar, St. Helena? (? indicates based on specimens with no rhizome data)
HAB. By forest streams, usually in swampy areas, rock faces above rivers, *Croton–Celtis* forest (near sea level); 1200–1900 m

SYN. *Aspidium gueinzianum* Mett., Farngatt. IV: 83 (1858) (as *gueintziana*)
    Lastrea gueinziana (Mett.) Moore, Ind. Fil.: 93 (1858) (as *gueintziana*)
    Nephrodium gueinzianum (Mett.) Hieron. in E.J. 28: 34 (1900) (as *gueintzianum*)
    N. prolixum sensu Peter, F.D.-O.A.: 56 (1929), *non* Bak.
    N. pallidivenium sensu Peter, F.D.-O.A. 1: 58 91929), *non* Bak.
    Thelypteris gueinziana (Mett.) Schelpe in J.S. Afr. Bot. 31: 262, 264 (1965) & F.Z., Pterid: 194 (1970) & Expl. Hydrobiol. Bassin L. Bangweolo & Luapula 8 (3) Ptérid.: 80 (1973) & C.F.A., Pterid.: 154 (1977); Schelpe & Anthony, F.S.A., Pterid.: 217 (1986); Burrows, S. Afr. Ferns: 266, t. 44/7, illustr. 63/271, 271a & b (1990)

NOTE. *Faden* 68/831 (Kenya, Chania R.) has pinnae 25 × 3 cm and lobes 17 × 5.5 mm. These exceptional measurements have not been included in the description.

8. **Christella chaseana** (*Schelpe*) *Holttum* in J. S. Afr. Bot. 40: 148 (1974); Schelpe & Diniz, Fl. Moçamb., Pterid.: 211 (1979); Jacobsen, Ferns S. Afr.: 392, fig. 296 (1983); Schippers in Fern Gaz. 14: 196 (1993); Faden in U.K.W.F., ed. 2: 32 (1994). Type: Namibia, Otjiwarongo District, *Schelpe* 4791 (BOL, holo.; BM!, iso.)

Rhizome creeping, with brown lanceolate acuminate entire slightly pilose scales up to 11 mm long. Fronds spaced up to 1 cm apart, 0.35–1.5 m tall. Stipe pale brown 30–63 cm long, glabrous. Lamina narrowly elliptic to lanceolate in outline, deeply bipinnatifid, 30–70 cm long, 15–40 cm wide, acuminate with a deeply pinnatifid terminal segment; sometimes up to 3 pairs of basal pinnae reduced and deflexed; middle pinnae narrowly oblong, 8–20 cm long, 1.8–2.3 cm wide, acuminate, pinnatifid into narrowly oblong lobes 5–10 mm long, 2–3.5 mm wide, sometimes falcate, obtuse; veins entirely free or contiguous, meeting in the sinus membrane, occasionally ± anastomosing. Indusia conspicuous, usually with dense long white hairs.

UGANDA. Toro District: 3 km N of Kichwamba, 23 Sept. 1969, *Faden* 69/1247!
KENYA. Trans-Nzoia District: 7.5 km NE of Kitale, Koitobos R., Sandum's Bridge, 12 June 1971, *Faden et al.* 71/456!; Teita District: Mbololo Hill, Mraru Ridge, 9 Apr. 1971, *Faden et al.* 71/243! & Taveta, Msau R. valley, 18 May 1985, *Kabuye et al.* T.H.E. 631!
TANZANIA. Arusha District: near Arusha, Sokon, 2 Nov. 1955, *Milne-Redhead & Taylor* 7026!; Mpanda District: Kapapa, forest beyond U.T.C. sawmill, 19 Sept. 1970, *Richards* 25980!; Iringa District: 38 km SE of Iringa, Dabaga Highlands, Kilolo, 10 Feb. 1962, *Polhill & Paulo* 1430!
DISTR. **U** 2; **K** 3, 4?, 7; **T** 2, 4, 7, 8; Cameroon, Congo (Kinshasa), Angola, Zambia, Malawi, Zimbabwe, Namibia
HAB. Swampy streamsides with *Eulophia*, *Thelypteris confluens* etc. very often in riverine forest, mist forest with *Macaranga*, *Albizia*, *Newtonia* etc., also in essentially drier areas, steep rocky slopes with *Terminalia*, *Combretum*, *Acacia* etc., streamsides in *Brachystegia–Uapaca* woodland, bogs in rough bushy grassland and in cultivations by streams; 800–1900 m

SYN. *Dryopteris dentatus* var. sensu Chiov., Racc. Bot. Miss. Consolata Kenia: 140 (1935) quoad *Balbo* 621 pro parte, *non* (Forssk.) C.Chr.
    Thelypteris chaseana Schelpe in J. S. Afr. Bot. 31: 263 (1965) & F.Z., Pterid.: 194 (1970) & Expl. Hydrobiol. Bassin L. Bangweolo & Luapula 8 (3) Ptérid.: 80 (1973) & C.F.A., Pterid.: 155 (1977); Schelpe & Anthony, F.S.A., Pterid.: 215 (1986); Burrows, S. Afr. Ferns: 265, t. 44/4, illustr. 63/270, 270a & b (1990)

NOTE. Some of the information for this species is taken from sheets determined by Schelpe or Holttum but have field notes which do not describe the rhizome or have inconsistencies e.g. creeping but fronds tufted. *Richards* 7383 (Tanzania, Ufipa District, Sumbawanga, Kaka R.,) has been determined by Alston as *Thelypteris pseudogueintziana* (Bonap.) Alston, but I do not see how it differs from the above species. If they prove to be the same, then Bonaparte's name is much the older.

9. **Christella** sp. A

Rhizome unknown. Stipe-base unknown but stipe over 20 cm long, pubescent. Lamina 30–60 cm long, 16–21 cm wide; in one of fronds seen the lowest pinna is only moderately shorter than those above but in the other there are several much reduced pinnae, the lowest only about 2 cm long; largest pinnae narrowly oblong-lanceolate, up to 12 cm long, 1.5 cm wide, narrowly acuminate, pinnatifid into oblong lobes about 8 mm long, 2.5 mm wide, densely pubescent above and on costae and costules beneath and with dense very pale yellow small glands beneath; veins with lowest pair free or connivent in the membrane or a few with distinct excurrent veins. Indusium densely hairy and with some glands.

KENYA. Northern Frontier District: Mt Nyiru, 13 Dec. 1972, *Cameron* 142!
DISTR. **K** 1, ?4, ?6 (see note)
HAB. By forest stream; ± 2400 m

NOTE. Holttum named the Kew sheet of this without reservation as *C. hispidula* (Decne.) Holttum and he mentions that capitate orange hairs sometimes occur in his description of that species. Faden has annotated the EAH sheet as "*Thelypteris sp. aff. gueintziana* (Mett.) Schelpe or possibly a hybrid with *Cyclosorus dentatus*". I suspect it could be a distinct taxon but rhizome data are needed. The possibility of hybrids in *Christella* could only be resolved by study of populations in the field. *Faden* 69/915 (Kenya, Kiambu District: Thika, behind the Blue Posts Hotel, under the

Chania Falls, 1440 m, 3 Aug. 1969) is described as rhizome creeping, branched; the seven lowest pinnae are abruptly reduced, the lowest under 1 cm long; there are numerous sessile yellow glands on the lower surface of the pinnules and glands on the indusia. I had at first, because of the free veins and creeping rhizome, looked on it as an aberrant *C. chaseana* but it is probably conspecific with sp. A. *Glover & Oledonet* 4511 (Kenya, Masai District, Mt Suswa, inside Cave 6, beneath sunlight hole in roof, 1800 m, 22 Mar. 1964) is the same. Other material which I suspect belongs here is *Bally* in CM 7458 (Kenya, Kiambu District, Thika Falls). Several specimens collected by *Mrs Prescott Decie* s.n. (Kenya, Ondoni R., 950 m, 1926 & Nyeri, 1926) also seem best placed here. More field work is needed in this easily accessible area.

10. **Christella** sp. B

Rhizome and stipe-base unknown. Lamina 32 cm long, 16 cm wide, the lowest pair of pinnae only slightly reduced; largest pinnae narrowly lanceolate, 8 cm long, 1.2 cm wide, pinnatifid into oblong lobes about 5 mm long, 2.5 mm wide, fairly densely hairy above, with few long hairs on costules beneath; veins with lowest pair anastomosing to form a rounded areole and excurrent nerve to the sinus. Indusium densely hairy with very long hairs ± 1 mm long exceeding width of indusium.

TANZANIA. Lushoto District: Usambaras, *Buchwald* 429!
DISTR. **T** 3; not known elsewhere
HAB. Not known, probably forest

NOTE. The label bears the name *Asplenium sandersonii*, so obviously wrong, that the label may not belong. I have seen nothing which matches it.

HYBRIDS

Holttum mentions (J. S. Afr. Bot. 40: 142 (1974)) under species 3 that the type of *Thelypteris dentata* var. *buchananii* is probably a hybrid between *Christella dentata* and *Pneumatopteris afra*. Quite a number of East African sheets seem to be similar. The hybrid *Pneumatopteris afra* × *Christella dentata* has been given a name based on West African material and it is convenient to use it.

Chrismatopteris N.Quansah & D.S.Edwards in K.B. 41: 805 (1986)

× **Chrismatopteris holttumii** *N.Quansah & D.S.Edwards* in K.B. 41: 805, figs. 1–3 (1986). Type: Ghana, Central Region, Efutu, *Quansah* 123 (K!, holo.)

Morphologically intermediate between the parents with rhizomes short- to long-creeping and veins 7–9 pairs with lower 2–3 pairs anastomosing or one may be missing on one side of the excurrent vein. Sporangia sometimes setose.

UGANDA. Toro District: Kibale National Park, near Kanywara, 14 Oct. 1995, *Poulsen et al.* 1016!; Mengo District: Kampala, Namanve Swamp, June 1937, *Chandler* 1691! & Marukota, Mpanga Forest Reserve, 5 km E of Mpigi, 9 Sept. 1969, *Faden et al* 69/998!
KENYA. North Kavirondo District: Kakamega Forest, Kibiri Block, S side of Yala R., 21 Jan. 1970, *Faden et al.* 70/23!; Kwale District: Shimba Hills, Kurumuji and Mwacho Mwana R. Valleys, 2 Aug. 1970, *Faden & Evans* 70/422 B!
TANZANIA. Buha District: Kasakela Reserve, Melinda Stream, 19 Nov. 1962, *Verdcourt* 3375!; Mpanda District: Kungwe-Mahali peninsula, S of Pasagulu, Kabwe R., 7 Aug. 1959, *Harley* 919!; Ulanga District: near Mangula, Maganga, 1 Sept. 1960, *Haerdi* 589/0!
DISTR. **U** 2, 4; **K** 5, 7; **T** 4, 6, 7; Sierra Leone, Liberia, Ghana, Nigeria, Cameroon
HAB. Wet evergreen forest, often riparian, forest and thicket bordering swamps; 150–1600 m

SYN. *Cyclosorus* sp. sensu Haerdi in Acta Trop. Suppl. 8: 34 (1964)
     C. elatus vel. sp. aff. sensu Haerdi in Acta Trop. Suppl. 8: 33 (1964)
     C. sp. A; Faden in U.K.W.F. ed. 1: 54 (1974)
     Christella sp. A; Faden in U.K.W.F. ed. 2: 31 (1994)

NOTE. Holttum had annotated *Faden & Evans* 70/422 B in 1971 with a new name based on Shimba, but it was never published. He stated 'apparently nearest to *Nephrodium distans* Hook. of Madagascar' (i.e. *Christella distans* (Hook.) Holttum). Holttum suggested that *Thelypteris dentata* var. *buchananii* Schelpe in J. S. Afr. Bot. 31: 265, fig. 1d (1965). Type: Mozambique, *Schelpe* 5599 BOL (holo.) is a similar hybrid but is beyond the known range of *Pneumatopteris afra*; recently it has been treated as a species *Christella buchananii* (Schelpe) J.P.Roux. Similarly *Uhlig* 1090 (Tanzania, Kilimanjaro, forest above Kibosho, 2900 m, 1901) has venation associated with the hybrid but *P. afra* does not occur on Kilimanjaro. It might be a hybrid with another species of *Pneumatopteris*. Ballard noticed the 2–3 pairs of united veins, and unhesitatingly annotated it as '*Dryopteris dentata*'. Study of material from Socotra and the type locality, Yemen throws doubt on some of the so called hybrids since two pairs of anastomosing veins sometimes occur. *Hepper & Wood* 6021 (Yemen, Wadi Dur, Udaya, 22 Oct. 1975) determined by Holttum as *C. dentata* definitely has two pairs of anastomosing veins at most sinuses also dense small glands and some hooked hairs beneath. The microfiche of Forsskål's type is not very clear but I think two pairs of anastomosing veins are sometimes present. There are undoubted true hybrids showing other characters such as slightly setose sporangia but at least some so-called hybrids could be mere forms of *C. dentata*. Detailed field studies and spore studies are needed; of the list of possible hybrids cited by Holttum (J. S. Afr. Bot. 40: 158 (1974)) he says "some of these have abortive spores". Viane in B.S.B.B. 118: 49 (1985) has described a hybrid *Thelypteris* × *varievenulosa* from the Ivory Coast which he thinks is probably *T. afra* × *T. hispidula* (i.e. *Pneumatopteris afra* × *Christella hispidula* in Holttum's nomenclature).

5. STEGNOGRAMMA

Blume, Pl. Jav.: 172 (1828); K. Iwatsuki in Acta Phytotax. Geobot. 19: 112–126 (1963)

Rhizome short-creeping or erect; stipes densely hairy with unicellular (in African species) or septate hairs. Fronds simply pinnate, the basal pinnae not or little reduced. Pinnae subentire to deeply-lobed, the upper ones always adnate to the rachis and in some species only the lowest fully free but usually many free; terminal lobe triangular; spherical glands lacking. Veins free or with many anastomoses. Indusium absent; sporangia setose.

A pantropical genus with about 18 species; a single species in Africa.

Stegnogramma pozoi (*Lagasca*) *K.Iwatsuki* in Acta Phytotax. Geobot. 19: 124 (1963); Jacobsen, Ferns S. Afr. 393:, fig. 297a, b (1983); Schippers in Fern Gaz. 14: 197 (1993); Faden in U.K.W.F. ed. 2: 32 (1994). Type: Spain, Cantabria, *Pozo* s.n. (S-PA, ?holo.)

Rhizome about 3 mm diameter, erect with dark brown lanceolate ciliolate scales 1–5 mm long. Fronds tufted, 15–80 cm long, simply pinnate. Stipe 15–30 cm long, pale brown, slightly pubescent with minute hairs. Lamina lanceolate to narrowly elliptic in outline, 25–40 cm long, (3–)6–14 cm wide, bipinnatifid; pinnae 12–15 pairs, either all more or less adnate to the rachis or lower 4–8 pairs free or minutely stalked; longest pinnae narrowly oblong or attenuate, 6.5–7.5(–13) cm long, 1.2–1.6(–2) cm wide, incised about half-way to costa into rounded lobes 6–8 mm long, 3–5 mm wide, pubescent on both surfaces; veins ± 6 pairs, entirely free. Sori linear-oblong or narrowly elliptic-oblong, the veins, without indusium.

SYN. *Hemionitis pozoi* Lagasca, Gen. Sp. Pl.: 33 (1816)
     Leptogramma pozoi (Lagasca) Heywood in F.R. 64: 19 (1961)

var. **petiolata** (*Ching*) *Sledge* in Bull. Brit. Mus. Bot. 8: 49 (1981, Feb.); Holttum in Fl. Males. ser. II, 1: 542 (1981, Dec.)*. Type: Sri Lanka, *G. Wall* s.n. ("Wallich" of Ching) (PE, holo.)

Lower 4–8 pairs of pinnae free. Fig. 5, p. 21.

* Volume 1 (5) is dated March 1982 in publication dates on page (4) and December 1981 on rear cover.

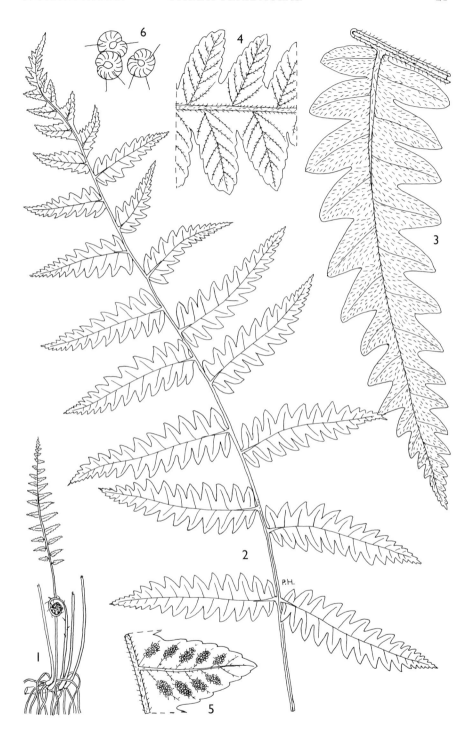

Fig. 5. *STEGNOGRAMMA POZOI VAR. PETIOLATA* — **1**, habit (not drawn to scale); **2**, frond × ²/₃; **3**, single pinna × 2; **4**, part of sterile pinna (under surface), enlarged; **5**, single fertile pinnule, enlarged; **6**, sori showing setae, enlarged. 1 from *Grimshaw* 93/882; 2–6 from *Fries & Fries* 1225. Drawn by Pat Halliday.

UGANDA. Mbale District: Mt Elgon, 24 Dec. 1957, *Molesworth-Allen* 3680! & Sasa Trail, 19 Apr. 1997, *Wesche* 1298!

KENYA. Northern Frontier District; Mt Nyiru, Mbarta forest, 29 Mar. 1995, *Bytebier et al.* 48!; Fort Hall District: Kimakia Forest Station, 13 July 1969, *Faden & Evans* 69/898; Kericho District: 5 km NW of Kericho, Kimugung R., 10 June 1972, *Faden et al.* 72/296!

TANZANIA. Arusha District: Mt Meru, crater floor gorge, 19 Jan. 1967, *Vesey-FitzGerald* 5056!; Mpanda District: Mahali Mts, Sisaga, 29 Aug. 1958, *Newbould & Jefford* 1914!; Morogoro District: Uluguru Mts, NE slope of Bondwa, just below peak, 16 Feb. 1971, *Pócs et al.* 6403/J!

DISTR. **U** 3; **K** 1, 3–5, 7; **T** 2, 4, 6, 7; Cameroon, Bioko, Congo (Kinshasa), Sudan, Ethiopia, Malawi, Zimbabwe, South Africa; Comoro Is., Sri Lanka and Java

HAB. Montane moorland, *Podocarpus–Hagenia*–bamboo upland forest, mixed riverine forest, often by springs or streams but also rocky places; 1800–3400 m

SYN. *Polypodium tottum* Willd., Sp. Pl. ed. 4, 5: 201 (1810). Type: South Africa, Cape of Good Hope, Herb. Willd. 19697 (B, holo.), *non* Thunb. (1800)

   P. africanum Desv. in Mem. Soc. Linn. Paris 6: 239 (1827), *nom. nov.*

   Nephrodium tottum (Willd.) Diels in E. & P. Pf. 1 (4): 170 (1899); Hieron. in V.E. 2: 12, fig. 7 (1908); F.D.-O.A. 1: 55 (1929)

   Dryopteris africana (Desv.) C.Chr., Ind. Fil.: 250 (1905); Sim, Ferns S. Afr. ed. 2: 102, t. 23 (1915)

   Leptogramma pilosiuscula sensu Alston, Ferns W.T.A.: 63 (1959), *non* (Wikstr.) Alston sensu stricto

   L. petiolata Ching in Acta Phytotax. Sin. 8: 319 (1963)

   Thelypteris pozoi sensu Tardieu, Fl. Cameroun 3: 239, t. 36/3, 4 (1964); Schelpe, F.Z. Pterid.: 199, t. 55/G (1970); Schelpe & Anthony, F.S.A. Pterid.: 213, fig. 69/1 (1986); Burrows, S. Afr. Ferns: 260, t. 43/4, illustr. 64/262, 262a, b (1990), *non* (Lagasca) Morton sensu stricto

   Leptogramma pozoi sensu Faden in U.K.W.F. ed. 1: 54 (1974); Pic. Serm. in B.J.B.B. 53: 284 (1983), *non* (Lagasca) Heywood sensu stricto

NOTE. The African material appears closer to that from Sri Lanka, morphologically at least, than it does to var. *pozoi* from Spain, Madeira and the Azores but the varieties need further investigation; intermediates are common in East Africa, *Luke & Luke* 4786 (NE Mt Kenya, Rotundu, Kazita R., 25 Sept. 1997) is close to the typical variety. Other varieties extend the range of the species to India and Japan.

6. **THELYPTERIS**

Schmidel, Ic. Pl. ed. Keller: 45, t. 11, 13 (1763)

Rhizome slender, long-creeping, branched, always in very wet ground, with thin broad glabrous scales; fronds widely spaced. Lamina simply pinnate with deeply pinnatifid pinnae, the basal ones not or little reduced; veins all free, often forked, running to the margin. Flat thin scales and sometimes filamentous smaller ones on lower surface of the costae. Sori indusiate; sporangia sometimes with short capitate hairs.

One or two species in N temperate region and one scattered in various areas mostly south of the equator and widespread in tropical Africa extending to 7°N.

Thelypteris confluens (*Thunb.*) *Morton* in Contr. U.S. Nat. Herb. 38: 71 (1967); Schelpe, F.Z., Pterid.: 190, t. 55/E (1970) & in Expl. Hydrobiol. Bassin L. Bangweolo & Luapula 8 (3) Ptérid.: 78, fig. 24A (1973); Holttum in J. S. Afr. Bot. 40: 150 (1974); Schelpe, C.F.A., Pterid.: 155 (1977); Schelpe & Diniz, Fl. Moçamb., Pterid.: 212 (1979); Holttum in Fl. Males. Ser. II, 1: 377 (1981); Pic. Serm. in B.J.B.B. 53: 275 (1983); Jacobsen, Ferns S. Afr.: 395, fig. 298 (1983); Schelpe & Anthony, F.S.A., Pterid.: 211, fig. 71/1 (1986); Burrows, S. Afr. Ferns: 258, t. 43/1, illustr. 62/260, 260a & b (1990); Schippers in Fern Gaz. 14: 198 (1993); Faden in U.K.W.F. ed. 2: 32 (1994). Type: South Africa, Cape Peninsula, *Thunberg* s.n. (UPS, holo., seen by Schelpe)

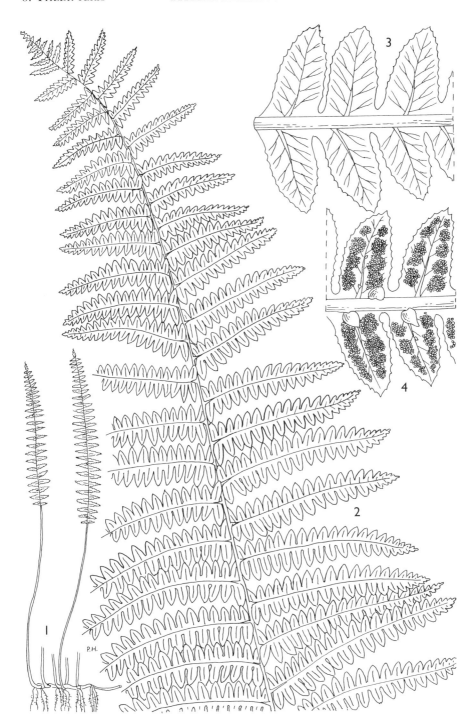

Fig. 6. *THELYPTERIS CONFLUENS* — **1**, habit (not drawn to scale); **2**, part of frond × ²/₃; **3**, part of pinna showing venation (diagrammatic); **4**, part of pinna showing scales and sori (diagrammatic). All from East African material. Drawn by Pat Halliday.

Rhizome widely creeping, up to 56 cm long, 2–3 mm in diameter with dark brown ovate somewhat undulate rhizome scales up to 2 mm long blackening with age. Fronds spaced up to 5 cm apart, 0.35–1.05 m tall. Stipes stramineous, black at the base, 15–50 cm long. Lamina lanceolate to broadly lanceolate in outline, 25–55 cm long, 21 cm wide, deeply bipinnatifid, acute at apex, the basal pinnae slightly reduced; pinnae narrowly oblong, 4–11 cm long, 0.6–2 cm wide, deeply lobed, the lobes oblong, 0.3–1.4 cm long, 3–5 mm wide, glabrous, glandular or thinly pilose beneath and with thin cinnamon membranous ovate scales along the costa beneath; costa grooved above, glabrous; veins in sterile fronds mostly forked, in fertile often simple. Indusium glabrous or with short marginal hairs. Fig. 6, p. 23.

UGANDA. Kigezi District: Kamuganguzi, 2 km on Kabale side of old rest house, swamp crossing, 28 July 1952, *Norman* 140!; Masaka District: Lake Nabugabo, July 1937, *Hancock & Chandler* 1755!; Mengo District: Kampala, King's Lake, 12 Dec. 1935, *Chandler & Hancock* 98!

KENYA. Trans-Nzoia District: 19 km SW of Kitale, Nov. 1959, *Tweedie* 1928!; Kiambu District: Ondiri Swamp, 2 Mar. 1952, *Rayner* 524!; Teita District: Taita Hills, Waruga, *Faden & Evans* 69/885!

TANZANIA. Arusha District: Ngurdoto National Park, Tululusie, 25 Oct. 1965, *Greenway & Kanuri* 12207!; Lushoto District: 6 km NE of Lushoto, Mkusi, 22 Apr. 1953, *Drummond & Hemsley* 2187!; Songea District: about 32 km E of Songea, R. Mukira, 26 Dec. 1955, *Milne-Redhead & Taylor* 7753!

DISTR. **U** 2, 4; **K** 3–7; **T** 1–4, 7, 8; Nigeria, Congo (Kinshasa), Rwanda, Burundi, Sudan, Ethiopia, Angola, Zambia, Malawi, Mozambique, Zimbabwe, Botswana, Namibia, South Africa; Madagascar, S India, N Thailand, Laos, N Sumatra, New Guinea, Australia (Queensland and Victoria) and New Zealand

HAB. Swampy areas, quaking *Sphagnum* bogs, *Typha, Papyrus, Cyperus* marshes but sometimes in seasonally drier areas, vleis banks and ditches, *Syzygium* forest; (0–)900–2400 m (see note)

SYN. *Pteris confluens* Thunb., Prodr. Fl. Cap.: 171 (1800)
    Aspidium thelypteris Sw. var. *squamigerum* Schltd., Adumbr. Fil. Prom. B. Spei: 23, t. 11 (1825). Type: South Africa, Cape Province, Hex R., *Mund & Maire* s.n. (?HAL, holo.)
    Dryopteris thelypteris (L.) Gray var. *squamigera* (Schltd.) C.Chr., Ind. Fil.: 297 (1905)
    Nephrodium thelypteris sensu Hieron. in V.E. 2: 10 (1908); Peter F.D.-OA. 1: 56 (1929), *non* (L.) Strempel
    Dryopteris thelypteris sensu Sim, Ferns S. Afr. ed. 2: 101, t. 16 (1915), *non* (L.)
    Thelypteris palustris Schott var. *squamigera* (Schltd.) Weath. in Contr. Gray Herb. n.s. 73: 40 (1924); Tardieu in Mém. I.F.A.N. 28: 119 (1953)
    Nephrodium totta sensu Peter, F.D.-O.A. 1: 55 (1929) quoad *Peter* 8812 & 37697, *non* (Willd.) Diels
    Thelypteris squamigera (Schltd.) Ching in Bull. Fan. Mem. Inst. Bot. 6: 329 (1936) (as "*squamulosa*" sphalm); Tardieu, Fl. Cameroun 3: 243 (1964)

NOTE. Holst 2567 (Tanzania, Amboni, June 1893) is from a much lower altitude (probably ± sea level) than any other East Africa specimen seen. It is curious it has not been collected again from such a well known area. Only one specimen seen for Nigeria, *Bowden* 71 (Sardauna Province, Mambila Plateau, valley near Maya–Ndaga, 6 Apr. 1970).

7. CYCLOSORUS

Link, Hort. Berol. 2: 128 (1833); Holttum in Bothalia 19: 27 (1971)

Rhizome long-creeping with rather broad scales with mostly marginal hairs. Lamina bipinnatifid with lower pinnae not reduced and with flat clathrate scales on the lower surface of the costae but these can be obscure or absent. Basal veins anastomosing with a long excurrent vein to the sinus. Spherical red glands sometimes present on costules and veins. Sori indusiate; indusia glabrous to pilose; stalks of sporangia with long hairs with terminal red glands.

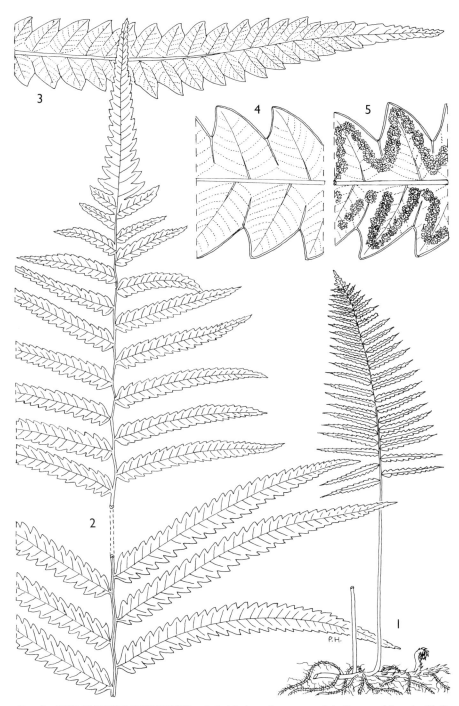

Fig. 7. *CYCLOSORUS INTERRUPTUS* — **1**, habit (not drawn to scale); **2**, part of frond × ²⁄₃; **3**, part of pinna × 2 ; **4**, part of sterile pinna showing venation (diagrammatic); **5**, part of fertile pinna showing sori (diagrammatic). All from *Milne-Redhead & Taylor* 7090 & *Beesley* 83. Drawn by Pat Halliday.

A pantropical genus with probably 3 species according to Holttum but I have found it impossible to use his key satisfactorily. I have also found it difficult to use Alston's key who also recognised 3 species from West Africa but used different characters. I recognise only one species and was therefore reassured to find that Jacobsen (Ferns S. Afr.: 395 (1983)) and A.R. Smith (Fl. Mesamer. 1: 181 (1995)) had also done so (see note).

Cyclosorus interruptus (*Willd.*) *H.Ito* in Bot. Mag. Tokyo 51: 714 (1937); Holttum in J. S. Afr. Bot. 40: 152 (1974); Schelpe & Diniz, Fl. Moçamb., Pterid.: 214 (1979); Holttum, Fl. Males. ser II, 1:386 (1981); Pic. Serm. in B.J.B.B. 53:280 (1983); Jacobsen, Ferns S. Afr.: 395, fig 299 (1983); Faden in U.K.W.F. ed. 2: 32 (1994). Type: S India, *Klein* in *Herb Willd.* 19770 (B-W, holo., seen by Holttum)

Rhizome long-creeping, 5–6 mm wide, with sparse black narrowly ovate acute entire scales up to 2 mm long. Fronds spaced up to 12(–15) cm apart, 0.7–2.4 m tall. Stipe pale brown, dark at base, 45–90 cm tall, glabrous. Lamina oblong-lanceolate, 30–85 cm long, 25–30 cm wide, bipinnatifid; pinnae narrowly oblong, 10–21 cm long, 1–2.5 cm wide, attenuate, deeply to ± shallowly lobed; lobes ovate, quadrate or narrowly oblong, 6–17 mm long, 4–7 mm wide, entirely glabrous to sparsely or densely pubescent beneath and often with distinct red glands and clathrate scales on costae and costulae beneath very evident or sparse or often absent; veins finely raised beneath. Sori round at first and close but usually eventually coalescing to form a u-shaped line around the sinuses in a very characteristic manner; indusium glabrous to densely pilose and sometimes with glands.

var. **interruptus**

Pinnae narrower, mostly less deeply cut to about ½ way into more ovate or triangular lobes, glabrous to densely hairy beneath and often with glands. Fig. 7, p. 26.

UGANDA. Karamoja District: Pian Co., Nakipiripirit, July 1965, *Wilson* 1698!; Kigezi District: Kamugangui, 1.6 km from Kabale side of swamp crossing, 28 July 1952, *Norman* 141!; Mengo District: Kampala, King's Lake, 5 Dec. 1935, *Chandler & Hancock* 100!
KENYA. Meru District: Nyambeni Hills, Maua, 31 May 1969, *Faden et al.* 69/652!; Masai District: Amboseli Reserve, 11 Sept. 1954, *Bally* 9871!; Kwale District: Mteza R., 1927, *Gardner* F.D. 1455!
TANZANIA. Moshi District: Arusha Chini, 16 km S of Moshi, Jan 1965, *Beesley* 83!; Tanga District: 13 km SW of Muhesa, Bombwera, 19 Nov. 1955, *Milne-Redhead & Taylor* 7090!; Rufiji District: Mafia I., near Ngombeni-Mafia, 16 Oct. 1965, *Cameron* 3!; Pemba: Mtangatwani bridge, 10 Oct. 1929, *Vaughan* 770!
DISTR. U 1–4; K 4, 6, 7; T 1–8; Z, P; widespread in tropical Africa, West Africa to Congo (Kinshasa), Sudan, Ethiopia to South Africa; tropical and subtropical Asia extending to Micronesia, Polynesia, Hawaii etc., and America and West Indies
HAB. *Papyrus* and *Cyperus latifolius* swamps, *Sphagnum* and *Miscanthidium* bogs, riverine bushland, seepage areas in *Brachystegia–Julbernardia* woodland, drainage ditches, sometimes persisting in marshy valleys in sisal plantations; 6–2100 m

SYN. *Pteris interrupta* Willd. in Phytographia 1: 13, t. 10, f. 1 (1794); Holttum in Amer. Fern J. 63: 81 (1973), *non* Willd. (1810)
   Polypodium tottum* Thunb., Prodr. Pl. Cap.: 172 (1800). Type: South Africa, Cape, Worcester Division, Brandvlei, *Thunberg* s.n. (UPS, holo., seen by Schelpe)
   Aspidium goggilodus** Schkuhr, Krypt. Gew. 1: 193, t. 33c (1809). Type: Guyana, Essequebo, collector ? (HAL, holo.)
   Nephrodium unitum sensu R.Br., Prodr. Fl. New Holland: 148 (1810); Peter, F.D.-O.A. 1: 57 (1929) quoad *Peter* 7838 & 18187, *non Polypodium unitum* L. sensu stricto

* Thunberg used the epithet *tottus* adjectivally for several plants but no such word appears to exist in Latin. I do not know what the derivation is.
** There is a perfectly good Greek word γογγυλωδης but it has nearly always been written gongylodes the change probably starting with C.F.W. Meyer, Prim. Fl. Essequiboensis: 230 (1818).

Cyclosorus goggilodus (Schkuhr) Link, Hort. Berol. 2: 128 (1833); Tardieu in Mém. I.F.A.N. 28: 26, t. 2/5–6 (1953); Alston, Ferns W.T.A.: 62 (1959) (as '*gongylodes*'); Tardieu, Fl. Cameroon 3: 246 (1964) (as '*gongylodes*'); Schippers in Fern Gaz. 14: 197 (1993) (as '*gongylodes*').
Dryopteris goggilodus (Schkuhr) O.Ktze, Rev. Gen. Pl. 2:811 (1891); Sim, Ferns S. Afr. ed. 2: 97, t. 13 (1915) (as '*gongylodes*')
Thelypteris totta (Thunb.) Schelpe in J. S. Afr. Bot. 29: 91 (1963) & F.Z. Pterid.: 198, t. 55/F (1970) & Expl. Hydrobiol. Bassin L. Bangweolo & Luapula 8(3), Ptérid.: 81 (1973)
T. interrupta (Willd.) K.Iwatsuki in J. Jap. Bot. 38: 314 (1963); Schelpe, C.F.A., Pterid.: 157 (1977); Schelpe & Anthony, F.S.A., Pterid.: 209, fig. 70 (1986); Burrows, S. Afr. Ferns: 259, t. 43. 2, illustr. 64/261, 261 a & b (1990); A.R. Smith in Fl. Mesoameric. 1: 181 (1995)

var. **striatus** (*Schumach.*) *Verdc.* **comb. nov**. Type: Dahomey, Ouidah [Whydah*], *Isert* s.n. (C, holo.; microfiche!)

Pinnae wider, more deeply cut to $\frac{3}{4}$ or more of way down to costa into more oblong more falcate lobes, rounded to often minutely denticulate at the apex, mostly ± glabrous beneath.

UGANDA. Kigezi District: illegible, 3 Aug. 1936, *A.S. Thomas* 2070!; Busoga District: 5 km SW of Kityerara Forest Station, Kityerara Port, 15 June 1953, *Wood* 758!; Mengo District: Kiagwe, Namanve Swamp, Mar. 1932, *Eggeling* 249 in F.D. 492!
KENYA. Central Kavirondo District: Lake Victoria, Maboko I., 25 Dec. 1939, *Mrs Hornby* S1088! (see note)
TANZANIA. Bukoba District: Nyakato, Oct. 1935, *Gillman* 418!.
DISTR. **U** 2–4; **K** 5; **T** 1; widespread in West Africa from Gambia to Angola, Bioko, Annobon, Chad, Congo (Brazzaville), Central African Republic, Congo (Kinshasa); recorded from Malawi and Zambia but material is close to *C. interruptus* sensu stricto
HAB. *Papyrus* swamps, saturated ground at edge of rivers etc.; 1100–1800 m

SYN. *Aspidium striatum* Schumach., Beskriv. Guin. Pl.; 456 (1827) & in K. Dansk. Vidensk. Selsk. 4: 230 (1829)
Polypodium pallidivenium Hook., Sp. Fil. 5: 8 (1963). Type: Sierra Leone, R. Bagroo, *Mann* 909 (K!, holo.)
Nephrodium pallidivenium (Hook.) Bak., Syn. Fil: 290 (1868); F.D.-O.A. 1: 58 (1929)
Cyclosorus striatus (Schumach.) Ching in Bull. Fan Mem. Inst. Biol. Bot. 10: 249 (1941); Tardieu in Mém. I.F.A.N. 28: 124, t. 22/1, 2 (1953); Alston, Ferns W.T.A: 62 (1959); Tardieu, Fl. Cameroun 3: 247 (1964) & Fl. Gabon 8: 151, t. 24/6–8 (1964); Faden in U.K.W.F. ed 2: 32 (1994)
Thelypteris striata (Schumach.) Schelpe in J. S. Afr. Bot. 31: 268 (1965) & F.Z., Pterid.: 199 (1970) & Expl. Hydrobiol. Bassin L. Bangweolo & Luapula 8 (3) Ptérid.: 82 (1973) & C.F.A., Pterid., 158 (1977)

NOTE. *Cyclosorus molundensis* (Brause) Pic.Serm. in Webbia 32: 77 (1977) (Type: Cameroon, Mulundu, *Mildbraed* 4382 (B, holo.; BM!, iso.) is very close to var. *striatus*, and Tardieu in Not. Syst. Paris 14: 345 (1953) has treated it as a variety of *C. striatus*. Pichi Sermolli states the pinnae are more stipitate (but they are slightly so in all specimens of this genus) and more deeply cut. Other material is more distinctive e.g. *Mildbraed* 8789a from Cameroon, Dengdeng which has the pinna lobes lanceolate, up to 3.2 cm long, 4.5 mm wide with broad sinuses. Similar specimens occur in Sierra Leone and Congo-Kinshasa and fide Pichi Sermolli in Central African Republic and Burundi. I have not therefore formally sunk it. *Hornby* S1088 cited above shows the very rounded sinuses of this taxon.

GENERAL NOTE. Holttum keeps *C. striatus*, *C. tottus* and *C. interrruptus* separate as follows:

1. Pinnae lobed $\frac{3}{4}$ or more toward costa often 2.5–3 cm wide, costules beneath with abundant very smooth scales . . . . *C. striatus*
   Pinnae less deeply lobed, rarely more than 2 cm wide; lower surface of costule glabrous or bearing acicular and/or capitate hairs, often also spherical glands 2
2. Pinnae quite hairless and eglandular beneath, very firm . . . . *C. tottus*
   Pinnae variously hairy (often densely) and glandular beneath, usually thinner . . . . *C. interruptus*

* Schumacher states 'In damp places at Whydah and Aquapim' but the specimen is from Whydah.

One can of course roughly sort out material according to this but there are very many intermediates and it makes no geographical sense of any sort. Holttum specifically says he is recognising three species for convenience but the large amount of material now available renders it now more inconvenient to try and do so.

8. AMPELOPTERIS

Kunze in Bot. Zeit. 6: 114 (1848)

Rhizome creeping, sometimes with non-peltate black scales. Fronds tufted, with indefinite apical growth and bearing many buds on the rachis from which new plants form freely. Lamina simply pinnate, the pinnae not deeply lobed; veins almost all anastomosing to form an excurrent vein, only a few residual ones free and running to the margin. Sori round or somewhat elongate without indusium; capitate paraphyses present.

A monotypic genus distributed throughout the tropics and subtropics of the Old World.

Ampelopteris prolifera (*Retz*) *Copeland*, Gen. Fil: 144 (1947); Tardieu, Fl. Madag. 5(1): 300 (1958); Alston, Ferns W.T.A.: 63 (1959); Haerdi in Acta Tropica, Suppl. 8: 32 (1964); Schelpe, F.Z., Pterid.: 200, t. 56 (1970) & Expl. Hydrobiol. Bassin L Bangweolo & Luapula 8 (3) Ptérid.: 82, fig. 25 (1973); Holttum in J. S. Afr. Bot. 40: 153 (1974); Schelpe, C.F.A., Pterid.: 160, t. 28 (1977); Schelpe & Diniz, Fl. Moçamb., Pterid.: 215, t.15 (1979); Holttum, Fl. Males. ser. II, 1: 387, fig. 7/d–f (1981); Pic. Serm. in B.J.B.B. 53: 284 (1983); Jacobsen, Ferns S. Afr.: 398, fig. 300 (1983); Schelpe & Anthony, F.S.A., Pterid. 220, fig. 74 (1986); Burrows, S. Afr. Ferns: 270, t. 45/3, illustr. 62, 276, 276 a & b (1990); Schippers in Fern Gaz. 14: 196 (1993); Faden in U.K.W.F. ed. 2: 32 (1994). Type*: S India, *Koenig* s.n. (LD, holo.)

Rhizome stout, up to 1 cm in diameter, with black triangular entire acuminate scales up to 2 mm long but these are sometimes absent. Fronds up to 1.2–1.4 m tall. Stipe up to 40 cm long. Lamina oblong or narrowly trianglar in outline, up to 1 m long, 26 cm wide, with basal pinnae hardly reduced; upper pinnae reduced towards the apex. Pinnae oblong-lanceolate to linear-oblong, 3–15 cm long, 0.8–2 cm wide, acute at the apex, truncate at the base, sessile or petiolate, very shallowly lobed or no more than crenate, glabrous; terminal pinnae ± lobed at base; veins up to 8 pairs with at least 5 pairs anastomosing. Sori coalescing to form a ± zig-zag line along the pinna curving round the lobes. Fig. 8, p. 29.

KENYA. Embu/Kitui District: along the R. Tana, *Faden* s.n.
TANZANIA. Moshi District: Kikafu R. bridge, 15 Apr. 1968, *Greenway & Kanuri* 13459!; Ulanga District: near Kiberege, Oct. 1960, *Haerdi* 612/0; Rungwe District: Kiwira (Kibila), 3 Nov. 1911, *Stolz* 953!; Zanzibar: Mwera Swamp, 19 Aug. 1960, *Fanshawe* 2686!
DISTR. **K** 4, **T** 2, 6, 7; **Z**; Senegal, Guinea, Cameroon, Congo (Kinshasa), Burundi, Angola, Zambia, Malawi, Mozambique, Zimbabwe, South Africa; also Madagascar and Mascarene Is.; widespread in tropics and subtropics of Old World to New Caledonia, New Guinea and Australia
HAB. Marshy places along rivers with sedges etc., trailing in water between rocks of waterfalls, riverine woodland; 0–1100 m

SYN. *Hemionitis prolifera* Retz., Obs. Bot. 6: 38 (1791)
      Meniscium proliferum (Retz.) Sw., Syn. Fil.:19, 207 (1806); Hook., Sec. Cent. Ferns, t. 15 pro parte excl. small complete plant and fig. 2 (1861)
      Nephrodium proliferum (Retz.) Keys., Pol. Cyath. Hb. Bung.: 49 (1873); Hieron. in V.E. 2: 12, fig. 8 (1908)

* Holttum gives GOET but Retzius's types are at Lund. Fischer (K.B. 1932: 75 (1932)) saw this one there.

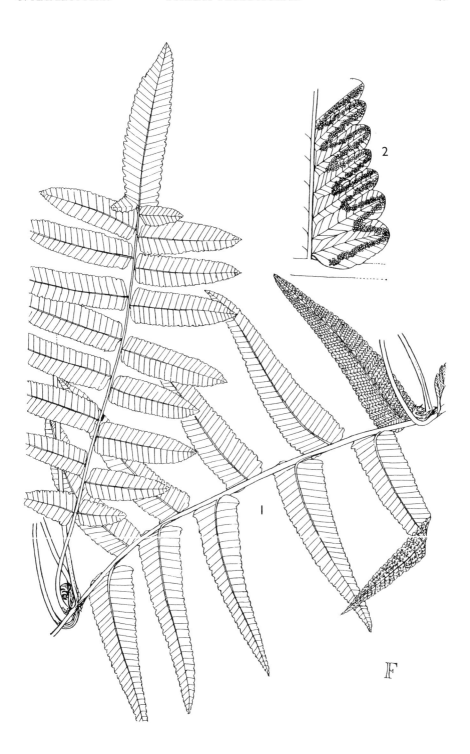

FIG. 8. *AMPELOPTERIS PROLIFERA* — **1**, part of frond × ²/₃; **2**, part of fertile pinna × 4. Both from *Welwitsch* s.n. Reproduced with permission from F.Z.

Dryopteris prolifera (Retz.) C.Chr., Ind. Fil.: 286 (1905); Sim, Ferns S. Afr. ed. 2: 99, t. 14 (1915)
Cyclosorus proliferus (Retz.) Tardieu & C.Chr. in Not. Syst. 14: 346 (1952); Tardieu, Mem.
    I.F.A.N. 28: 128, t. 22/11–13 (1953)

9. MENISORUS

Alston in Bol. Soc. Brot. sér. 2, 30: 20 (1956)

Rhizome short, erect with few brown ovate scales. Fronds tufted. Lamina simply
pinnate with basal pinnae not reduced. Pinnae ± 10–17, alternate or ± opposite,
narrowly lanceolate, serrate or crenate but not deeply lobed, glabrous; apical pinna
simple, with gemma at base. Lateral nerves reaching tips of teeth; tertiary veins in
2–3 pairs, anastomosing. Sori in ± 1–2 lines ± parallel to costa, without indusia.
Sporangia not setose.

A monotypic genus widely distributed in Tropical Africa.

Menisorus pauciflorus (*Hook.*) *Alston* in Bol. Soc. Brot. sér 2: 20 (1956) & Ferns
W.T.A.: 63 (1959); Holttum in J. S. Afr. Bot. 40: 154 (1974); Schelpe, C.F.A., Pterid.: 160,
t. 27 (1972); Pic. Serm. in B.J.B.B. 53: 284 (1983); Schippers in Fern Gaz.: 197 (1993).
Type: Equatorial Guinea/Gabon, Crystal Mts (Sierra del Crystal), *Mann* 1672 (K!, syn.)

Fronds up to 15, tufted, 20–85 cm tall; stipe 10–30 cm long with broadly ovate
glabrous basal scales. Lamina (13–)30–50 cm long, (7–)13–25 cm wide; pinnae 5–15 cm
long, 0.5–1.5 cm wide, attenuate at apex, cuneate at the base. Young plant from
gemma often remaining on plant for some time. Fig. 9, p. 31.

UGANDA. Ankole District: Kasyoha-Kitomi Forest Reserve, NE of Kyambura R., 9 June 1994,
    Poulsen et al. 555!; Kigezi District: Ishasha Gorge, 5 km SW of Kirima, 21 Sept. 1969, *Faden et
    al.* 69/1188! & Kayonza Forest, in stream running into Ishasha R., June 1952, *H.D. van
    Someren* 930!, 944!, 945!, & 946!
TANZANIA. Morogoro District: NE Uluguru Mts, 14 May 1933, *Schlieben* 3932!; Iringa District:
    Udzungwa Mts, above Sanje waterfalls, 5 Nov. 1998, *de Boer et al.* 773! & same area,
    Mwanihana Forest, 8 Sept. 1984, *D.W. Thomas* 3655!
DISTR. U 2; **T** 6, 7; Nigeria, Cameroon, Congo (Kinshasa), Central African Republic, Equatorial
    Guinea/Gabon, Sudan, Angola
HAB. On rocks in and along rivers in riverine forest, also lower montane forest and mist forest;
    950–1500 m

SYN. *Meniscium pauciflorum* Hook., Sp. Fil. 5:164 (1864)
    Dryopteris pauciflora (Hook.) C.Chr., Index Fil.: 283 (1905); Reimers in N.B.G.B. 11: 393 (1933)
    Polypodium prionodes C.H.Wright in K.B. 1906: 253 (1906). Type: Uganda, Ankole District:
        W Ankole Forest, *Dawe* 369 (K!, syn.)
    Dryopteris pauciflora (Hook.) C.Chr. var. *pruinosa* Reimers in N.B.G.B. 12: 79 (1934). Type:
        Tanzania, Morogoro District: NE Uluguru Mts, *Schlieben* 3932 (B, holo.; BM!, iso.)

10. PNEUMATOPTERIS

Nakai in Bot. Mag. Tokyo 47: 179 (1953); Holttum in Blumea 21: 293–325 (1973)

Rhizome usually erect but creeping in a few species (some African), the scales
usually thin and broad with few marginal hairs. Fronds sometimes covered with
mucilage when very young. Lamina sometime ± pustular when dry, with lower (often
many) pairs of pinnae gradually or abruptly reduced; aerophores at bases of pinnae
often swollen and distinct; veins usually anastomosing (always in African species) or
free in about a dozen species. Sporangia sometimes bearing capitate hairs, setiferous
in one African species. Indusium present or absent.

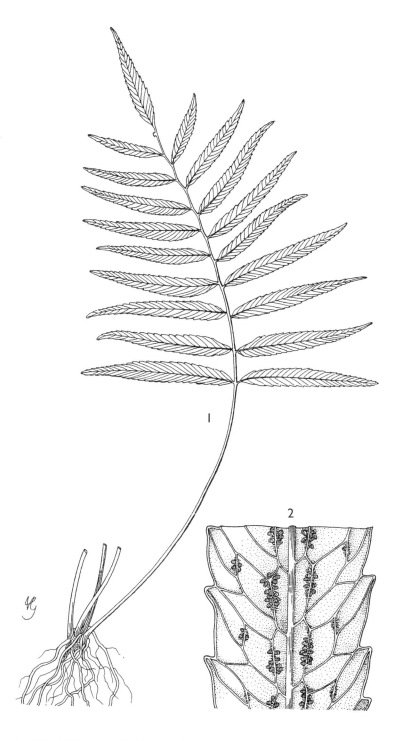

FIG. 9. *MENISORUS PAUCIFLORUS* — **1**, habit × ²⁄₃; **2**, part of fertile pinna × 6. Both from *Price & Evans* 123. Taken with permission from Conspectus Florae Angolensis, Pteridophyta, Tab. XXVII (1977). Drawn by Victoria Goaman.

About 75 species from W Africa to Hawaii and Queensland.

1. **Pneumatopteris unita** (*Kunze*) *Holttum* in Blumea 21: 304 (1973) & in J. S. Afr. Bot. 40: 155 (1974); Schelpe & Diniz, Fl. Moçamb.: 216 (1979); Jacobsen, Ferns S. Afr: 399, fig. 301 (1983); Schippers in Fern Gaz. 14: 197 (1973); Faden in U.K.W.F. ed. 2: 32 (1994). Type: South Africa, Durban (Port Natal), *Gueinzius* s.n. (LZ, holo.†; HBG, iso, BOL, photo.)

Rhizome erect. Fronds tufted, often large, 0.9–3.6 m tall, covered with mucilage when young and coiled; gemmae present on rachis at apex of frond. Stipe pale brown, 22–75 cm long, glabrous. Lamina lanceolate in outline, 0.6–1.5 m long, 20–50 cm wide, simply pinnate, the basal pinnae hardly reduced; basal pinnae auricled; pinnae linear-oblong, 11.5–28 cm long, 1.5–3.3 cm wide, rather shallowly lobed to $^1/_4$–$^1/_3$ to costa; lobes oblong, somewhat falcate, 4–10 mm long, 3–7 mm wide, glabrous, save for scattered short hairs on costa etc. above; about 4 pairs of veins anastomosing below the sinus either in a membrane or extending from the sinus or below it. Indusium absent. Fig. 10, p. 33.

UGANDA. Toro District: Kazingo, Bwamba Pass, 11 Jan. 1932, *Hazel* 172!; Ankole District: 5–7 km NW of sawmill W of Rubuzigye, 19 Sept. 1969, *Lye et al.* 4129! & W Ankole, Buhweju, Kaakara–Mpija, 9 July 1988, *Rwaburindore* 2652!
KENYA. Naivasha District: S Kinangop, Sasamua Dam Pipeline Road, 11 Dec. 1960, *Verdcourt et al.* 3033!; Meru District: Nyambeni Tea Estate, 8 Oct. 1960, *Verdcourt & Polhill* 2930!; Kericho District: SW Mau Forest, 16 km SSE of Kericho, along Kiptigat/Chepkuisi R., 12 June 1972, *Faden et al.* 72/347!
TANZANIA. Moshi District: Kilimanjaro, Mweka route, 27 July 1968, *Bigger* 2036!; Lushoto District: W Usambaras, Shagayu Forest Reserve, 2.5 km SW of Shagein peak, 22 Oct. 1986, *Pocs* 86207/A!; Rungwe District: Livingstone Mts, Isalola R. above Bumbigi, 5 Mar. 1991, *Gereau & Kagambo* 4226!
DISTR. **U** 2; **K** 3–5, 7; **T** 2, 3, 5–7; Liberia, Ghana, Cameroon, Congo (Kinshasa), Sudan, Malawi, Mozambique, Zimbabwe, South Africa (Natal); Madagascar
HAB. Evergreen rain, swamp and bamboo forests, roadsides and forest edges; (700– fide Schippers) 1450–2500 m

SYN. *Gymnogramma unita* Kunze in Linnaea 18: 115 (1844)
   Goniopteris madagascariensis Fée, Gen. Fil.: 251 (1852). Type: Madagascar, *Goudot* s.n. (RB, holo.?)
   G. patens Fée, Gen. Fil.: 253 (1852). Type: South Africa, Natal, *Gueinzius* s.n. (RB, holo?)
   G. silvatica Pappe & Rawson, Syn. Fil. Afr. Austr.: 39 (1858), *nom. illegit.* Type as for *Pneumatopteris unita*
   Polypodium unitum (Kunze) Hook., Spec. Fil. 5: 5 (1863), *non* L. *nom. illegit.*
   Dryopteris silvatica (Pappe & Rawson) C.Chr., Ind. Fil.: 292 (1905); Sim. Ferns S. Afr. ed. 2: 100, t. 15 (1915), *nom. illegit.*

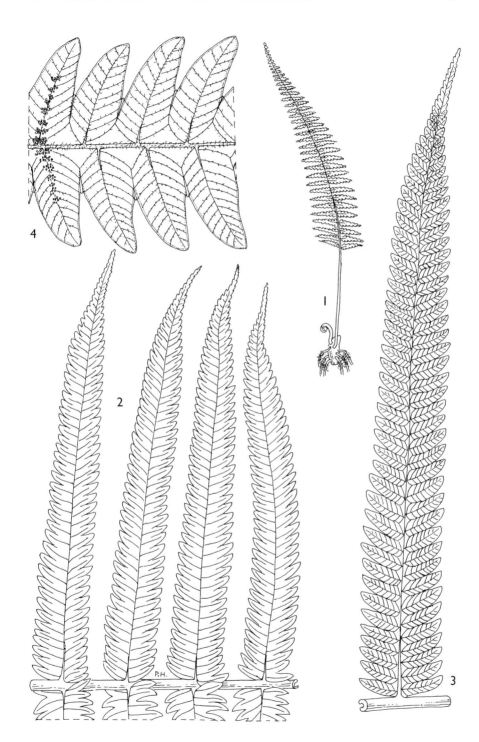

FIG. 10. *PNEUMATOPTERIS UNITA* — **1**, habit (not drawn to scale); **2**, part of frond × ²/₃; **3**, pinna × 1 ; **4**, part of pinna (fertile and sterile) showing venation (diagrammatic). Drawn from *Verdcourt* 3986 & *Faden* 69/681. Drawn by Pat Halliday.

Dryopteris albidipilosa R.Bonap., Notes Ptérid. 15: 9 (1924). Type: Kenya, Fort Hall District: near Thika, *Mearns* 1123 (P, holo.; K!, photo.; BM!, US, iso.)*

D. silvatica (Pappe & Rawson) C.Chr. var. *amplectens* C.Chr. in N.B.G.B. 9: 176 (1924). Type: Kenya, Mt Kenya, W Kenya Forest Station, *R.E. & T.C.E. Fries* 1217 (?S, holo.; BM!, iso.)

Cyclosorus patens (Fée) Copel., Gen. Fil.: 143 (1947); Alston, Ferns W.T.A.: 62 (1959); Tardieu, Fl. Cameroun 3: 250, t. 36/1–2 (1964)

C. albidipilosus (R.Bonap.) Tardieu in Not. Syst. 14: 345 (1952) & Mém. I.F.A.N. 28: 123, t. 21/3 (1953)

Thelypteris madagascariensis (Fée) Schelpe in J. S. Afr. Bot. 31: 267 (1965) & F.Z. Pterid.: 196 (1970); Schelpe & Anthony, F.S.A,: Pterid.: 209, fig. 69/2 (1986); Burrows, S. Afr. Ferns: 262, t. 43/6, illustr. 62/265, 265 a, b, c (1990)

2. **Pneumatopteris blastophora** (*Alston*) *Holttum* in J. S. Afr. Bot. 40: 156 (1974). Type: Nigeria, Ogoja Province, R. Ata, Koloishe, *Savory & Keay* FHI 25062 (BM, holo.; FHI, iso.)

Rhizome shortly creeping, ± 5 cm in diameter. Fronds laxly tufted or shortly spaced, 1.1–2.5 m tall. Stipe straw-coloured, 30–80 cm long. Lamina ± oblong in outline, 0.6–1.1 m long, 13–40 cm wide, the lower pinnae not reduced, the upper segment resembling the pinnae and with a gemma at its base; pinnae linear-oblong, 15–23 cm long, 2.5–3.5 cm wide, narrowly attenuate at the apex, truncate or rounded at the base, crenate; lobes up to 4 mm long, 6 mm wide; veins 8 with 4–5 anastomosing with zigzag excurrent vein. Rachis and sometimes costa shortly hairy above, glabrous or minutely hairy beneath. Indusium absent.

UGANDA. Kigezi District: 7 km SW of Kirima, along the Ishasha R., 21 Sept. 1969, *Faden et al.* 69/1229! & Bwindi National Park, Kayonza, near the Ishasha R., 12 Mar. 1995, *Poulsen et al.* 783!; Mengo District: East Mengo, 10 km W of Lugazi, Ssezzibwa Falls, 8 Sept. 1969, *Faden & Evans* 69/979!

DISTR. **U** 2, 4; Liberia, Ghana, Nigeria, Cameroon, Bioko

HAB. Swampy places on floor of evergreen forest, riverine forest; 1200–1300 m

SYN. *Cyclosorus blastophorus* Alston in Bol. Soc. Bot. sér., 2, 30: 12 (1956) & Ferns W.T.A.: 62 (1959)

C. patens sensu Alston, Ferns W.T.A.: 62 (1959) pro parte, *non* (Fée) Copeland

Abacopteris letouzeyi Tardieu** in Not. Syst. 16: 202, t.1/5, 6 (1960). Type: Cameroon, S of Mt Koupé, *Letouzey* 405 (P, holo.)

Thelypteris blastophora (Alston) Reed in Phytogia 17: 264 (1968)

NOTE. I am not convinced about the specific distinctness of this and *P. unita*. Alston confused them and Holttum (on determination labels admits to intermediates). I have followed him until more field work is done.

3. **Pneumatopteris afra** (*Christ*) *Holttum* in Blumea 21: 306 (1973) & in J. S. Afr. Bot. 40: 157 (1974); Schippers in Fern Gaz. 14: 197 (1993). Type: Central African Republic, Haut-Oubangui, near Ouaka, *Chevalier* 5799 (P, lecto.; K!, iso.)

Rhizome long-creeping, ± 7 mm in diameter. Fronds spaced ± 5 cm apart, 1–2 m tall. Stipe 35–54 cm long with narrow hairy scales at base. Lamina ovate-oblong, 30–60 cm long, 25–34 cm wide, simply pinnate, the basal 2–3 pairs of pinnae sometimes abruptly reduced and strongly auricled, the lowest only 1–2 cm long;

---

* I am grateful to R. Faden for looking at the material in US and identifying it. Later I discovered an isotype at BM. There is another Mearns specimen at BM also labelled 1123 but which is a different species *Cyclosorus interruptus* (Willd.) H.Ito. Schelpe (adnot.) points out this does not match Bonaparte's description.

** This is not mentioned in Fl. Cameroun, Ptérid. but is mentioned by Benl in Acta Bot. Barcinonensia 38: 60 (1988). The sori are in 6 lines ± parallel to the costa of the pinnae and appear most distinctive according to a photograph of a similar Bioko specimen sent to Kew by Benl. He saw the Letouzey type.

aerophores white and slightly swollen in living material; largest pinnae linear-oblong, 11–24 cm long, 2–5 cm wide, long acuminate at the apex, broadly cuneate at the base, crenate, the lobes ovate-oblong, 3–7 mm long, 2.5–5 mm wide; lower surface of costa etc. shortly hairy; veins in 10–12 pairs with 4–6 pairs anastomosing. Indusia present, hairy. Sporangia usually with 2–3 setae.

UGANDA. Bunyoro District: Budongo Forest Reserve, near Sonso R., 16 Sept. 1995, *Poulsen et al.* 965!; Toro District: Kibanga, Bwamba, 21 Nov. 1935, *A.S. Thomas* 1500!; Mengo District: East Mengo, 10 km W of Lugazi, Ssezzibwa Falls, 8 Sept. 1969, *Faden & Evans* 69/977!

TANZANIA. Buha District: Gombe Stream Reserve, Kakombe Valley, 25 Dec. 1963, *Pirozynski* 88!; Morogoro District: Uluguru Mts, Morningside, 28 Nov. 1934, *E.M. Bruce* 10!; Iringa District: Udzungwa Mountain National Park, Sonjo-Mwanihana route, 8 Nov. 1997, *P.A. & W.R.Q. Luke* 5008A!

DISTR. **U** 2, 4; **T** 4, 6, 7; Guinea, Sierra Leone, Liberia, Ivory Coast, Ghana, Nigeria, Cameroon, Gabon, Bioko, Central African Republic, Congo (Kinshasa), Burundi, Angola

HAB. Evergreen forest swamps, *Khaya* forest, streamside forest; 650–1500 m

SYN. *Dryopteris afra* Christ in Bull. Soc. Bot. France 55 [Mem. 8b]107 (1908)
    Cyclosorus afer (Christ) Ching in Bull. Fan Mem. Inst. Biol. Bot. 10: 242 (1941); Alston, Ferns W.T.A.: 63 (1959); Tardieu, Fl. Cameroun 3: 250, t. 37/1–3 (1964) & Fl. Gabon 8: 152, t. 24/1–3 (1964)
    C. oppositifolius sensu Tardieu, Mém. I.F.A.N. 28: 128, t. 21/4–6 (1953) pro parte, *non* (Hook. f.) Tardieu sensu stricto
    Thelypteris afra (Christ) Reed in Phytologia 17: 258 (1968); Schelpe, C.F.A. Pterid.: 158 (1977)

NOTE. For hybrids of this with *Christella dentata* see × *Chrismatopteris* p. 19. *Taylor* 3232A (Uganda, Toro District: Mpanga Forest, 25 Jan. 1935, almost lacks sporangial setae.

4. **Pneumatopteris usambarensis** *Holttum* in Blumea 21: 312 (1973) & in J. S. Afr. Bot. 40: 159 (1974); Schippers in Fern Gaz. 14: 197 (1993). Type: Tanzania, E Usambara Mts, Amani area, Marvera, *Faden et al.* 70/293 (EA, holo., one frond on 7 sheets[*]; K!, iso.)

Rhizome shortly creeping or erect. Fronds tufted, 1–2 m tall. Stipe 30–45 cm long with small adpressed scales at base. Lamina oblong, 0.65–1.1m long, 26–52 cm wide, simply pinnate, the lower 3 pairs, 5–8 mm long and 4 preceding pairs reduced; longest pinnae linear-oblong, 11.5–30 cm long, 1.5–2.6 cm wide, long acuminate at apex, with few hairs above, glabrous beneath or rarely with few hairs, lobed $^1/_3$–$^2/_3$ to costa; lobes semicircular to ovate-oblong, 2–7 mm long, 4–4.5 mm wide, subtruncate to round at apex, toothed or entire; veins 8–10 pairs, 2 basal pairs anastomosing. Indusia glabrous or slightly hairy, rarely conspicuously hairy; sporangia not setose.

KENYA. Teita District: Mbololo Hill, Sept.–Oct. 1938, *Joanna* in CM 9035!

TANZANIA. Lushoto District: Kwamkoro to Kihuhwi, 16 Dec. 1936, *Greenway* 4794! & W Usambaras, Mazumbai Forest Reserve, 25 Apr. 1975, *Hepper & Field* 5145!; Iringa District: Mwanihana Forest Reserve, above Sanje Village, 10 Oct. 1984, *D.W. Thomas* 3920!

DISTR. **K** 7; **T** 3, 6, 7; not known elsewhere

HAB. Evergreen forest; 300–1500 m

SYN. *Nephrodium truncatum* sensu Peter, F.D.-O.A. 1: 59 (1929), *non* (Poir.) Presl

* Strictly speaking this should be one holotype and 6 isotypes but since it was a single frond deliberately collected as a single specimen I consider it absurd not to look on it as a single holotype. The sheets should be joined in some way to make one entity.

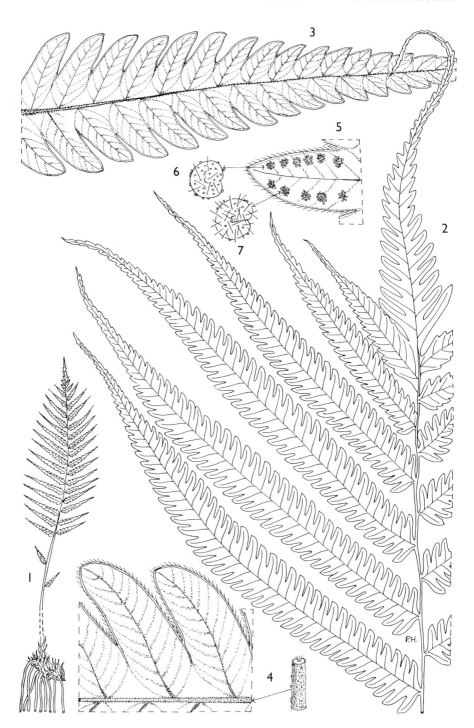

Fig. 11. *AMPHINEURON OPULENTUM* — **1**, habit (not drawn to scale); **2**, part of frond × ²/₃;
3, part of single pinna × 2; **4**, detail of part of pinna (diagrammatic); **5**, fertile pinnule
(diagrammatic); **6**, **7**, indusia showing variable indumentum (enlarged and diagrammatic).
All from *Faden et al.* 70/352 & Thomas 3743. Drawn by Pat Halliday.

11. AMPHINEURON

Holttum in Blumea 19: 45 (1971)

Rhizome short- to long-creeping or erect, with narrow setiferous scales. Fronds tufted or spaced. Stipe minutely hairy, scaly at the base. Lamina pinnate into usually deeply lobed pinnae; in some species with 1–3 pairs of short irregular pinnae at the base; veins simple, the basal ones either free or passing to the margin separately or connivent in the sinus-membrane or anastomosing to form an excurrent vein but sometimes variable in a single frond; sinus-membrane often ending in a tooth; lower surface of pinna with acicular and glandular hairs or glands. Indusium usually present, hairy or glandular.

About 12 species from East Africa to SE Asia, Malesia, Queensland and Pacific to Tahiti.

Amphineuron opulentum (*Kaulf.*) *Holttum* in Blumea 19: 45 (1971) & J. S. Afr. Bot. 40: 161 (1974) & in Blumea 23: 212 (1977); Schelpe & Diniz, Fl. Moçamb., Pterid.: 218 (1979); Holttum in Fl. Males. ser. II, 1: 548, fig. 19/b, c (1981); Jacobsen, Ferns S. Afr.: 401 (1983); Schippers in Fern Gaz. 14: 196 (1993). Type: Guam, *Chamisso* s.n. (LE, holo.)

Rhizome erect or short-creeping, branched, with dark brown attenuate minutely ciliate scales ± 9 mm long. Fronds tufted, 0.9–0.5 m tall. Stipe 40–70 cm long, minutely pubescent and with scales at base. Lamina broadly oblong-ovate, 0.8–1 m long, 60 cm wide, bipinnatifid, the basal pinnae not or rarely reduced; largest pinnae narrowly oblong, 25–35(–40) cm long, 2.5–2.9(–3.5) cm wide, acuminate, hairy above on costa, with short hairs and yellow glands beneath (particularly along veins), deeply lobed; lobes narrowly oblong, oblique, slightly falcate, 1.2 cm long, 4–5 mm wide; veins 8–10 pairs, the basal pair passing to sides of sinus-membrane or connivent or uniting below it forming a short excurrent vein. Sori coalescing to form lines parallel to lobe margins; indusia with small yellow glands on edges and often hairy (in extra-African material). Fig. 11, p. 36.

KENYA. Mombasa, *Wakefield* s.n.!
TANZANIA. Pangani District: Ntakuja, Sakura Mwera, 22 May 1956, *Tanner* 2836!; Morogoro District: Kimboza Forest Reserve, 4 July 1970, *Faden et al.* 70/352!; Iringa District: Mwanihana Forest Reserve, above Sanje, 25 Sept. 1984. *D.W. Thomas* 3743!; Pemba: km 8, Wete road, Semiwani, 5 Aug. 1929, *Vaughan* 456!
DISTR. **K** 7; **T** 3, 6, 7; **P**; Mozambique; Chagos Archipelago, Seychelles, tropical Asia and Malesia to N Queensland, eastwards to the Society Is. and Marquesas
HAB. Evergreen forest on slopes, streamside in deep shade, *Pandanus* swamp with numerous calcareous outcrops; near sea level–950m

SYN. *Aspidium opulentum* Kaulf., Enum. Fil. Chamisso: 238 (1824)
    A. extensum Bl., Enum. Pl. Jav.: 156 (1828). Type: Indonesia, Java, Pulo Pinang, *Blume* s.n. (L, holo.)
    Nephrodium wakefieldii Bak. in Ann. Bot. 5: 326 (1891); Hieron. in P.O.A. C.: 85 (1895). Type: Kenya, Mombasa, *Wakefield* s.n. (K!, lecto.) (4 sheets at K of which one annotated as type by Holttum)
    Dryopteris wakefieldii (Bak.) C.Chr., Ind. Fil.: 301 (1905)
    Thelypteris extensa (Bl.) Morton in Amer. Fern J. 49: 113 (1959); Schelpe, F.Z. Pterid.: 193 (1970)
    T. opulenta (Kaulf.) Fosberg in Smith. Contrib. Bot. 8: 3 (1972)

12. **SPHAEROSTEPHANOS**

J.Sm. in Hook, Gen. Fil.: t. 24 (1839); Holttum in Blumea 19: 29 (1971)

Rhizome predominantly erect or short-creeping but in a few species long-creeping or scandent with usually thin narrow scales with acicular hairs. Lamina pinnate, usually with abruptly or gradually reduced basal pinnae, rarely with no basal pinnae reduced; pinnae mostly lobed or crenate, rarely subentire; veins anastomosing or less often just meeting at the sinus or ending above it; sessile spherical yellow glands usually present on parts of the pinnae or on indusia or sporangia but sometimes quite lacking; acicular hairs nearly always present on costae and costules. Indusia usually present; sporangia sometimes with yellow glands and occasionally setose.

A large genus of over 150 species throughout Malesia, also in mainland tropical Asia, Australia and Pacific to Tahiti, Mascarene Is., Madagascar and São Tomé.

Rhizome erect; pinnae less coriaceous, crenate and with sparse
  glands above . . . .   1. *S. arbuscula*
Rhizome long-creeping; pinnae more coriaceous, more distinctly
  lobed, without glands above . . . .   2. *S. unitus*

1. **Sphaerostephanos arbuscula** * (*Willd.*) *Holttum* in J. S. Afr. Bot. 40: 165 (1974). Type: Mauritius, *Bory de St. Vincent* in Herb. Willd. 19763 (B-W, holo., seen by Holttum)

Rhizome erect to 30 cm or more tall with thick layer of roots and dark brown lanceolate distinctly hairy scales 3–4 mm long. Fronds tufted 0.5–1.5 m tall. Stipe 5–10 cm long, densely shortly adpressed hairy. Lamina narrowly oblong, 35–70 cm long, 12–28 cm wide, bipinnatifid, with 5–10 pairs of lower pinnae gradually reduced, all auricled at acroscopic base or not; largest pinnae linear-lanceolate, 6–17 cm long, 7–16 mm wide, acuminate to a rather blunt apex, distinctly auricled at base, usually distinctly crenate-serrate for 1–1.5 mm but sometimes only slightly so, lower surface densely hairy and with spherical glands also some glands on upper surface; lobes shortly obliquely ovate, up to 3 mm long and wide; veins 4–6 pairs, $1\frac{1}{2}$ pairs anastomosing next to the short sinus-membrane. Sori forming 3–4 lines on each side ± parallel to pinna costa; indusium with short hairs and glands; sporangia with yellow glands. Fig. 12, p. 39.

SYN. *Aspidium arbuscula* Willd., Sp. Pl. ed. 4, 5: 233 (1810)
    Thelypteris arbuscula (Willd.) K.Iwats. in Acta Phytotax. Geobot. 21: 170 (1965)

subsp. **africanus** *Holttum* in J. S. Afr. Bot. 40: 165 (1974); Schippers in Fern Gaz. 14: 197 (1993). Type: Kenya, Kwale District: Shimba Hills, *Drummond & Hemsley* 1203 (K!, holo.)

Pinnae larger, 7.5–16 cm long, 1–1.7 cm wide, not auricled at the acroscopic base but only slightly dilated on both sides at base; upper surface of pinnae only rather sparsely glandular.

KENYA. Kwale District: Shimba Hills, 14 km SW of Kwale, Pengo Forest, 11 Feb. 1953, *Drummond & Hemsley* 1203**! & Kurumuji and Mwacho R. valleys, 2 Aug. 1970, *Faden & Evans* 70/413!
TANZANIA. Lushoto District: E Usambara Mts, Nderema, Hunga R., 30 Mar. 1974, *Faden & Faden* 34/366!; Morogoro District: Nguru S Forest Reserve, behind Mhonda Mission, 6 Feb. 1971, *Mabberley & Pócs* 699!; Iringa District: Udzungwa Mt National Park, Mwaya–Mwanihana route, 5 Nov. 1997, *Luke & Luke* 4903!

---

  * To be treated as a noun in apposition.
** One sheet of this bearing only a rhizome had been annotated by Holttum as *Pneumatopteris usambarensis*, clearly a labelling error.

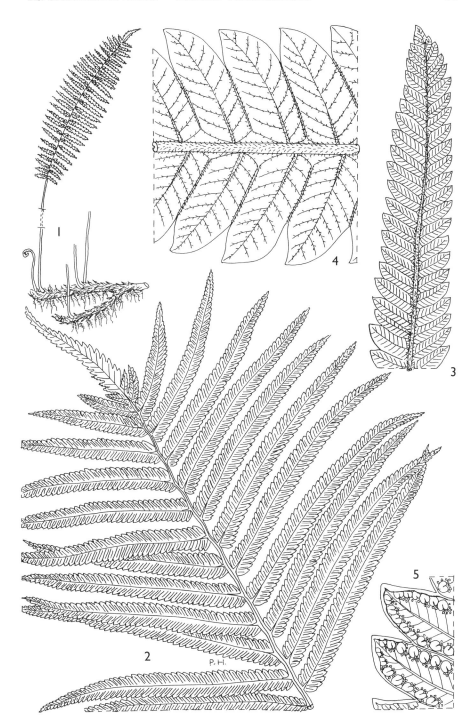

FIG. 12. *SPHAEROSTEPHANOS UNITUS* — **1**, habit (not drawn to scale); **2**, part of frond ×
$^2/_3$; **3**, part of single pinna (upper surface) × 2 ; **4**, detail of undersurface of part of pinna
(diagrammatic); **5**, part of fertile pinna showing undersurface (diagrammatic). All from
Stolz 1101, *Faden* 74/383 & *Richards* 15315. Drawn by Pat Halliday.

Distr. **K** 7; **T** 3, 6, 7; not known elsewhere

Hab. Streamside and riverine fringing forest of *Breonadia* and other Rubiaceae, lowland *Allanblackia* rainforest and intermediate rainforest; sometimes on rocky river banks; 150–1000 m

Note. I have seen no material with fronds quite as large as Holttum's measurements. The typical subspecies occurs in Mauritius, Réunion, Madagascar, Sri Lanka and S India.

2. **Sphaerostephanos unitus** (*L.*) *Holttum* in J. S. Afr. Bot. 40: 165 (1974) & in Kalikasan 4: 63 (1975) & Fl. Males. II, 1: 477 (1981); Schippers in Fern Gaz.: 197 (1993). Type: locality unknown, collector unknown (LINN, lecto., seen by Holttum)

Rhizome long-creeping, 5–7 mm in diameter with pale brown lanceolate-triangular acute glabrous scales 2–3 mm long. Fronds widely spaced, coriaceous, 1.2 m tall. Stipe 10–58 cm long; lamina oblong to ovate, 30–60 cm long, 20–25 cm wide, bipinnatifid with 6–12 lower pairs of pinnae reduced, the smallest about 1 cm long and numerous aerophores beneath; largest pinnae linear-lanceolate, 10–17 cm long, 0.8–1.5 cm wide, narrowly attenuate to the apex, hairy beneath on costa and lower surface with many orange-brown glossy glands (can be lacking or restricted to veins and costulae in non-African varieties), lobed for about $^{1}/_{3}$ into ovate-oblong lobes, 2–4 mm long, 2.5 mm wide; veins 8–10 pairs, $1^{1}/_{2}$ pairs anastomosing with next 2–4 pairs passing to sides of the sinus membrane. Sori ± 7 on each side of costule; indusia ± glabrous or with few hairs and glands; sporangia with glands.

var. **unitus**; Holttum in Fl. Males. II. I: 478 (1981)

Rather large glossy brownish glands present beneath pinnae, on and between veins and on costules.

Tanzania. Lushoto District: Kwamkoro, 16 Aug. 1961, *Richards* 15315!; Morogoro District: Uluguru Mts, Morogoro to Morningside, 2 April 1974, *Faden & Faden* 74/383!; Iringa District: Udzungwa Mt National Park, above Camp Site 2, 10 Nov. 1997, *Luke & Luke* 5110!

Distr. **T** 3, 6, 7; Malawi; Seychelles, Mascarene Is., Madagascar, Sri Lanka, S India, Assam to Burma, Indochina and Philippines, throughout Malesia to New Guinea, Guam and N Queensland

Hab. Rain forest, damp ground and roadside banks; 800–1400 m

Syn. *Polypodium unitum* L., Syst. Nat. ed. 10, 2: 1326 (1759) (excl. syn.)
       Nephrodium leuconeuron Fée, Gen. Fil.: 306, t. 18c/3 (1852). Type: Réunion, *Olivier* s.n.*
       Thelypteris unita (L.) Morton in Amer. Fern J. 49: 113 (1959)
       T. leuconeuron (Fée) Schelpe in J. S. Afr. Bot. 31: 266 (1965) excl. syn. *Nephrodium mauritianum* Fée

Note. The synonymy of this species is very extensive but not relevant to East Africa; it can be found in the Holttum references given above. The distribution is extended by two further varieties (differing in the presence or absence and positioning of the glands) to Borneo, Micronesia and Polynesia. From a purely nomenclatural point of view it should be noted that care is needed to sort out the synonyms based on *Polypodium unitum* L. and those based on *Gymnogramma unita* Kunze. There is confusion in both the literature and on herbarium labels.

* Many of Fée's types are at RB but this species is not listed by Windisch in Amer. Fern J. 72: 56–60 (1982)

ADDENDUM

As this part was going to press Andreas Hemp sent me a specimen of *Macrothelypteris torresiana* (Gaudich.) Ching collected in Tanzania, Morogoro District, above Morogoro near Morningside, road embankment near a water canal, 1200 m, Dec. 2004, *Hemp* 4195. This occurs in tropical Asia and the western Pacific Islands and is now widely naturalized in tropical America, but is otherwise known only from Natal in mainland Africa. It will doubtless spread rapidly. The fronds are bipinnate and it will key to *Pseudophegopteris*, but is immediately identifiable by the long stiff slender multicellular hairs up to 1.5 mm long on the undersurface. Such hairs are an exception in the family.

INDEX TO THELYPTERIDACEAE

New names validated in this part

Cyclosorus interruptus (*Willd.*) *H.Ito* var. **striatus** (*Schumach.*) *Verdc.*, **comb. et stat. nov**.

PLANTS PEOPLE
POSSIBILITIES

First published in 2006 by
Royal Botanic Gardens, Kew
Richmond, Surrey, TW9 3AB, UK
www.kew.org

ISBN 1 84246 150 8

British Library Cataloguing in Publication Data
A catalogue record for this book is available from the British Library

Design and typesetting by Margaret Newman,
Kew Publishing, Royal Botanic Gardens, Kew.

Printed in the UK by Hobbs the Printers

For information or to purchase all Kew titles please visit
www.kewbooks.com or email publishing@kew.org

All proceeds go to support Kew's work in saving the world's plants for life